Workplace Security Playbook

Workplace Security Playbook

The New Manager's Guide to Security Risk

Bob Hayes
Contributing Editor

ELSEVIER
AMSTERDAM • BOSTON • HEIDELBERG • LONDON
NEW YORK • OXFORD • PARIS • SAN DIEGO
SAN FRANCISCO • SINGAPORE • SYDNEY • TOKYO
Security Executive Council

Elsevier
The Boulevard, Langford Lane, Kidlington, Oxford, OX5 1GB, UK
225 Wyman Street, Waltham, MA 02451, USA

First published 2013

British Library Cataloguing in Publication Data
A catalogue record for this book is available from the British Library

Library of Congress Cataloging-in-Publication Data
A catalog record for this book is available from the Library of Congress

ISBN: 978-0-12-417245-6

For more publications in the Elsevier Risk Management and Security Collection, visit our website at store.elsevier.com/SecurityExecutiveCouncil

This book has been manufactured using Print On Demand technology. Each copy is produced to order and is limited to black ink. The online version of this book will show color figures where appropriate.

CONTENTS

EXECUTIVE SUMMARY

Creating guidelines for effectively managing workplace security involves a considerable amount of planning, effort, and cost. When employees lacking a security background find themselves in charge of the safety of company personnel or property, the development of a workplace security program becomes an even greater challenge. The *Workplace Security Playbook* provides the essentials these non-security employees will need to start filling in the pieces of prevention strategies that work for their organizations.

This playbook includes recommendations for implementing and maintaining a successful security program, and, where applicable, how to do so in conjunction with corporate security. It describes the basics of investigative procedures and resources, presents guidelines for performing inspections, discusses how to plan for local emergency situations and special circumstances, and helps build a reference source of procedural information and local contacts for a complete, A-to-Z security manual.

WHAT IS A PLAYBOOK?

A playbook is an excellent tool for the security or business leader who wants to develop, implement, enhance, or validate a specific aspect of a security or risk management program. Playbooks provide a detailed treatment of a security program or service that can be quickly and effectively applied to an immediate need within an organization. Playbooks define and present the essential elements most often used by successful practitioners. They provide a framework that a security professional can use to set up, manage, and communicate the program to stakeholders. They can also be used by non-security personnel who need an introduction and plan for action on a new security-related job responsibility. Playbooks are particularly useful for educators who are committed to providing current, relevant information and practices distilled from successful practitioners and programs that have a direct correlation to current security positions.

INTRODUCTION

Not all organizations have the luxury of a dedicated security team at every location. At times, employees without a background in security may need to make decisions related to the safety of their company's personnel and property. The *Workplace Security Playbook* was designed with the companies and sites that have other business functions coordinating security in mind. This playbook describes the basic concepts of security and provides guidelines that can be used to supplement an existing corporate security program or help build a new security program from the ground up.

The guidelines in this playbook are intended for any employee who is responsible for people, facilities, or assets. Whatever your role, these guidelines will become an important resource for managing security responsibilities on a daily basis. If you manage a facility that is a part of a larger corporate entity, the security guidelines you choose to implement should always be cleared with the corporate security team in advance or be in clear accordance with corporate security policy.

There is no one document that can be taken off the shelf and implemented in every business culture; however, it's possible to adapt the procedures and best practices discussed in this playbook, which has chapters categorized by issues, to create risk management guidelines that are customized to meet your unique needs. Although it includes some key elements that most companies have in common, this playbook does not cover sector-specific issues.

The *Workplace Security Playbook* should not be construed as a means to establish any legal standard of care or identify what reasonably prudent security precautions should be taken in any specific situation. The actions to be taken for individual situations will vary depending on the corporate culture and individual circumstances at the time. Ultimately, you must assess the situation, choose a response, and manage the consequences.

The material this playbook contains was devised over the course of twenty years, with updates as trends and risks changed.

CHAPTER 1

Security Performance Guidelines

Security is the responsibility of every manager. Just as business operations are constantly assessed, managers too must continually assess the organization's security risks. These risks are not static: they change and grow along with the business. New technology, increasing business competition, and social and cultural developments all contribute to new security risks and trends.

In this chapter, you'll find eight guidelines that present a course of action based on levels of performance, not specifications. In this way, these guidelines help assess the individual, changing business and security needs that you may face.

If your company has a corporate security department, look to them as a resource to help you identify the measures that will allow you to achieve your security goals and apply the performance guidelines included in this playbook in a manner that fits your unique situation.

This chapter doesn't describe all the security measures you can use to achieve the levels of security you need. For information about specific security tools and measures:

- Chapters 2 through 8 cover the basics for managers and coordinators, and recommend an ongoing program to handle security issues.
- Chapter 9 provides an outline for building your own reference source of local security information.

1.1 PERFORMANCE GUIDELINES VS. STANDARDS

The diversity of your company's operations and locations may make it impractical to employ standardized security technologies or procedures for all sites. To remedy this, you should identify areas of security that apply to all sites, and describe the basic levels of performance that will strengthen business operations while accommodating what resources are currently available.

The mission of security is to protect and develop company assets. In effect, this makes corporate security your business consultant, whose objective is cost-effective, workable security measures that match your business risks. The only way to achieve this objective is by helping you assess your needs on a case-by-case basis.

To address this diversity of needs, rather than issuing security standards this document provides performance guidelines. A standard, for example, might require all sites to install CCTV (closed circuit television) cameras at entrances to site perimeters, and to monitor the cameras 24 hours per day. That standard might be a good measure in many cases, but it may not be appropriate for your site for a host of reasons. You may need to consider the following questions:

- Do your operations really require that level of security?
- Can the cost of equipment be justified?
- Do you have the staff to implement that solution?
- Will you be located at that particular site long enough to get a return on your investment?
- What do you do if your site perimeter is actually a wall in a leased or shared office building?

Focus instead on performance guidelines. For example, a performance guideline for controlling access to your site starts with the assumption that it is important to control and monitor access, doing so by using appropriate measures.

1.2 THREE-STEP DECISION-MAKING PROCESS

The concept of performance guidelines depends on a business approach. The performance guideline of controlling access, for example, does not tell you how to do it. The option you choose—how you design, engineer, and implement that end result—is up to you. You'll find that the following three-step decision-making process is one you may already use for other facets of your business:

1. *Assess Your Needs*
   Assess both your overall and specific security needs, and evaluate them in light of your business operations. What is the probability of a risk becoming reality? How critical would this be to you and the company as a whole? What is your degree of vulnerability—the extent to which you cannot prevent or respond to the situation?

2. *Identify Your Results and Options*
   Once you have assessed your needs, go on to identify your end results and the measures you can use to achieve them. Do you want to reduce theft or vandalism, or limit your exposure to liability? These guidelines will help you identify your intended results, and then you can start to identify the specific options available to you.
3. *Choose the Right Solution*
   Select and implement the right security measures. The options you've identified may have trade-offs. Which of them can you support, both financially and operationally? What will best complement your overall security efforts and your business strategy and philosophy?

Working through this decision-making process is a fundamental first step toward implementing the eight performance guidelines described in this chapter.

1.3 PERFORMANCE GUIDELINES

The eight security performance guidelines discussed in this section will help you and your team:

- Identify the areas and levels of security needed;
- Integrate security into the environment and business operations;
- Control physical access to the site, buildings, and offices;
- Control access to proprietary information;
- Detect unauthorized access;
- Prepare for incidents and emergencies;
- Respond to incidents and emergencies; and
- Report incidents promptly and correctly.

1.3.1 Performance Guideline #1: Identify Areas and Levels of Security

This guideline focuses on four issues: the possible risks you face, the probability of specific incidents, how critical these incidents would be to you and the company, and your vulnerability in these areas. These points will help you identify the actual levels of security you need to establish.

The possibility of incidents depends on a number of factors, such as your location, type of business, and general security trends. Possible risks include natural disasters, theft, information loss, emergencies arising from accidents or intentional acts, and equipment failure.

The probability of incidents can be assessed by reviewing past incidents at your location, in your community, and at other locations. The nature of your operations can also indicate the probability of certain events.

Criticality is the impact that an incident would have. Would it pose a threat to personal safety? Would it affect a small amount of product, or would it stop or delay production for hours, weeks, or months? Would it affect the reputation of your business or location? Would it reflect negatively on the company?

Vulnerability is the measure of how easily an incident might occur. For example, how easy is it to bypass access controls, such as jumping a fence or falsifying an ID? How easy is it to throw a switch accidentally or to disregard security and safety precautions?

Assessing the level of security you need will help you identify and implement the measures that will remove or reduce your risks.

1.3.2 Performance Guideline #2: Design Security into Your Facility and Operations

Security should be designed into your facility and operations, not just added on. A simplistic breakdown of security might show three elements: physical barriers (gates and doors, for example), control systems (such as card access), and people (procedures and employee awareness). These elements support each other, and are the key components in designing and integrating security into your environment.

For example, you might choose a combination of typical security measures: procedures, access control systems, badges, keys, cameras, and alarms. But designing these components into your operation means you must consider several issues:

- Can these measures be implemented effectively?
- Will the systems and procedures complement the work environment and the efforts of your employees?
- Will the measures support each other, or are they merely redundant?
- Will the measures require increased security staffing?
- Will your investment in security technology offer a real improvement in the performance and cost-effectiveness of your security program?

Physical security measures should be integrated into a new facility at the planning and design stage, though appropriate security measures can also be retrofitted into an existing site. In other cases, engineering solutions may be secondary to your need to establish or reinforce procedural controls. Whatever your situation, designing and integrating security into your environment will deliver a more efficient, cost-effective, and workable system.

1.3.3 Performance Guideline #3: Control Physical Access to Your Site, Buildings, and Offices

In general, controlling employee and visitor access means that you question someone's authorization and either grant or deny access. There are many ways to accomplish this. Signs, barriers, and gates may control access onto a site. Card access systems, badges, and escort policies can effectively control access to and within a building. In some cases, a simple sign-in/sign-out procedure with a receptionist or security officer may be sufficient.

You may need to rely on these techniques for a variety of situations involving on-site employees, visiting employees, family members, vendors, and temporary or contract workers. You may also need to consider implementing measures for both regular work schedules and after business hours.

Whatever combination of measures you use, employee awareness and cooperation are fundamental to the success of all access control systems.

1.3.4 Performance Guideline #4: Control Access to Proprietary Information

Information security depends on preventing random, uncontrolled access to company information. Physical controls play an important part: controlling access into buildings and offices, locking desks and file cabinets, and otherwise physically securing documents. Data processing safeguards, passwords, document classification, and other procedural controls also help to handle sensitive information securely.

Information cannot always be physically confined and controlled. Conversations in public places, on the phone, or at trade shows and conventions also present risks to information security. For this reason, constant employee awareness of the need for information security is important.

1.3.5 Performance Guideline #5: Detect Unauthorized Access to the Site, Buildings, Offices, and Proprietary Information

Being able to detect unauthorized access to any business asset is a key security function. This includes both successful and unsuccessful attempts of unauthorized access. Access control systems, cameras, alarms, and door seals are some options, but any measure you use should provide an audit trail, such as automated exception reporting, sign-in/sign-out sheets, and computer log-on reports.

Detection methods serve as an early warning system and require continuous validation and testing.

1.3.6 Performance Guideline #6: Plan and Prepare for Security Incidents and Emergencies

No security program is fail-safe. Planning and preparing for the incidents and emergencies that are most likely to occur will help you quickly respond to a situation and control its effects.

There are various types of security incidents and emergencies to prepare for: unauthorized access, theft, injury, sabotage, equipment failure, environmental accidents, and natural disasters, to name a few. All of these present specific security issues, and you should plan for these possibilities. Identify the responsibilities and procedures for reporting and handling incidents and emergencies, and the internal and external resources that are available to help you. Determine how you will respond to the situation, control its effects, and resume any affected business operations.

1.3.7 Performance Guideline #7: Respond to Incidents and Emergencies with Prompt and Appropriate Actions

Responding effectively to an incident means you can quickly address the situation, contain its effects, and return to normal business operations as soon as possible. Incidents can be minor or they can escalate to emergency proportions. Your response might range from simply installing a new lock, to coordinating emergency assistance with public safety and law enforcement authorities.

No security measure is complete without the element of response: alarms need to be answered, thefts need to be investigated, and emergencies need to be managed. Responding in an appropriate way reduces the further loss of company assets and controls exposure to liability.

1.3.8 Performance Guideline #8: Promptly Report All Incidents, Crimes, and Policy Violations

If your company has a corporate security team, then this final guideline will apply to you. Reporting promptly to corporate security ensures that you receive the advice and help you need to handle incidents appropriately and prevent similar incidents from happening in the future. Reporting guidelines and procedures will be discussed in more detail in Chapter 4.

CHAPTER 2

Elements of a Facility Security Program

A comprehensive, ongoing security program will help you make sure that the security measures you select will remain a valid business investment. A security program pulls together the people, technology, and procedures that your security system relies on and integrates them with your business.

With a security program in place, you can monitor and maintain your operations and be sure that your options stay in step with your changing needs. The primary responsibility for security lies with the facility manager, but an effective security program begins with these building blocks:

1. Educating employees about their individual responsibilities.
2. Appointing a security coordinator.
3. Establishing a security committee.

2.1 EMPLOYEE AWARENESS

People are the key to a successful security program. All employees should be aware that security is their business, and a successful security program depends on their support.

2.1.1 Employee Safety

One of the major objectives of security is to provide a safe and secure working environment for your employees. They need to know and understand that their well-being is of primary importance to the company, and that security measures are being taken on their behalf.

2.1.2 Theft Protection

The majority of security-related incidents in the workplace involve people who are authorized to enter the facility. Only a small percentage of incidents, such as vandalism and burglary, arise from external sources.

For this reason, employee awareness also includes letting your employees know that security policies and procedures must be followed. Just as security exists to protect them, it also exists to protect your other assets—your products, property, and information.

2.1.3 Employee Responsibilities

Outline your employees' individual responsibilities in security. Their precautions and actions should include

- Promptly reporting suspicious activities or missing items;
- Checking their workplace and securing or locking up any valuable personal and company property, especially when it will be left unattended (during lunch and at the end of the day, for example);
- Securing sensitive or proprietary information, both on paper and on the computer;
- Following standard security procedures;
- Understanding and following emergency procedures;
- Escorting visitors; and
- Being aware of and identifying strangers in their areas, sorting out the legitimate business visitor from the unusual situation.

2.1.4 Management Support

Security objectives require the full support of management. Managers and supervisors can improve security by working together to identify security needs and to establish workable procedures that reduce risk and complement employee efforts.

2.1.5 Ongoing Awareness

Educating employees about security cannot be accomplished overnight. You will need to make an ongoing effort to sustain their awareness. You can reinforce your message by

- Discussing security issues at department or crew meetings;
- Holding periodic conferences or facility meetings;
- Including security articles in employee newsletters;
- Posting notices on bulletin boards;
- Coordinating special training and awareness programs through corporate security; and
- Providing security orientation to new employees, contractors, vendors, service people, and temporary employees.

Note: In Chapter 8, tools you can use to promote ongoing employee awareness of security are described in more depth.

2.2 SECURITY COORDINATOR

The security coordinator should be a key part of the management team who oversees security on an ongoing basis. The coordinator acts as the liaison between management, employees, and corporate security (if applicable) on security issues.

Depending on the size of your facility and the number and type of security concerns, security responsibilities may require a full-time coordinator. For smaller facilities, the coordinator can usually add these security functions to his or her current responsibilities.

2.2.1 Responsibilities

The major responsibilities of the security coordinator are listed below. Most of these functions are described elsewhere in these guidelines.

- Gather incident reports (see Chapter 4).
- Observe and report on security situations.
- Complete security surveys and validation tests (see Chapter 3).
- Provide local information for periodic management reviews.
- Serve as the liaison with local law enforcement personnel.
- Coordinate the services of local security companies (alarms and locks, for example).
- Coordinate employee identification cards, access control badge systems, and lock and key control programs.
- Develop access control procedures.
- Facilitate the purchase and installation of security systems or services.
- Contact corporate security for advice and clarification.
- Identify local training needs and coordinate programs with corporate security (see Chapter 8).
- Review and support corporate security programs, including the information security program.
- Play an active role in developing an emergency notification program (see Chapter 6).
- Develop a plan for handling demonstrations (see Chapter 6).
- Schedule site visits from corporate security and keep in regular contact for advice and clarification (if applicable).

To ensure the effectiveness of your security coordinator, upper management must give that person clear direction, as well as provide him or her with the authority and resources necessary to do the job.

2.3 SECURITY COMMITTEE

2.3.1 Responsibilities

Rather than take on the bulk of your company's security responsibilities, it's important to create a committee that addresses security concerns as a group. Often an existing committee, such as a management or operations committee, can also assume responsibility for security.

It's the responsibility of the committee to work with the security coordinator and facility manager, helping them to

- Identify and discuss security issues and concerns;
- Determine practical solutions to security needs;
- Adapt security performance guidelines to the facility's needs;
- Respond to the security coordinator's concerns about needs or security discrepancies; and
- Educate and train employees, making security part of everyone's responsibility.

If no appropriate committee exists that can assume responsibility at your facility, you should form one, using a management or operations committee as the model.

2.3.2 Selecting Members

Put together a committee that represents a variety of activities and levels within your facility. The members should either be the function heads or individuals who can represent each key function at the facility.

The functions represented should include:

- Management
- Warehousing
- Clerical
- Security
- Data processing
- Safety
- Production

- Quality
- Shipping and receiving
- Laboratory
- Administration
- Human resources

If there is more than one division at the facility, each division should be represented on the committee.

CHAPTER 3

Surveys and Validations

The security measures you invest in are not final solutions. The appropriateness of your security measures, as well as the effectiveness of your entire security program, must be continuously monitored and validated.

As discussed in Chapter 1, the levels and types of security you need will change due to many factors. Your current and future business operations are a primary indicator of your changing security needs. But the probability of incidents and your vulnerability to them can also depend on outside factors.

This chapter explains the surveys and validation tests that will help you address your changing security needs.

3.1 SECURITY SURVEYS

An in-depth security survey of your facility should be conducted at least annually. How frequently you perform a survey depends on a variety of factors that take into account your local environment, your operations, potential security issues, your security history, and your current level of protection.

This survey analysis of your site will consist of

- Meeting with the facility manager, security coordinator, and the function heads in key areas, such as management, sales/marketing, lab directors, production, shipping, warehousing, administration, human resources, and others;
- Touring the facility; and
- Completing a comprehensive assessment of your physical and procedural security needs. This assessment is used to evaluate the specialized needs of your site and operations.

After the survey is completed, have the security coordinator draft a set of security recommendations for areas that appear to need improvement. These recommendations may be simple, such as installing signs or locking doors, but whatever they consist of, they should also address both your security and business needs. Common recommendations address

- Ensuring employee safety;
- Protecting proprietary information;
- Strengthening access control;
- Forming a security committee;
- Validating your existing security measures;
- Planning for crises and disaster recovery; and
- Strengthening the resources you devote to security.

3.2 VALIDATION TESTING

A system of periodic validation testing is the quality control function for your entire security program: it shows you whether your security program is actually delivering the results you expect and meeting the performance guidelines outlined in Chapter 1. If it appears your security measures are not working, you can use the test results to help determine the cause of the failure and the corrective action you need to take.

When designing validation tests, keep in mind that they usually take less than 30 minutes to complete, allowing the tester to easily and thoroughly validate specific elements of your security program.

The various types of validation tests can include the following:

- You can validate your alarm systems by opening alarmed doors and evaluating the response. Did the alarm go off? Did the alarm company call the right people? Did your security officers and employees respond correctly?
- Check to see if any reports or papers that are marked “Restricted Information” have been left on unattended desks or in other unsecured areas.
- Try placing an extra pallet on a loading dock to test shipping and receiving controls. Are over-shipments being identified and handled according to procedures?
- Try to process a bogus transaction to test your financial controls, make a random check of employees to see if they have all the keys that are assigned to them, or have a visitor try to gain access without the required identification.

3.3 TYPES OF VALIDATIONS

Validations should cover a wide range of security checks. The following list shows you the types of systems you should be testing.

- Access controls, either manual or electronic, including systems and procedures for the site perimeter, building, and special security areas
- Data processing controls and information security program efforts
- Identification systems
- Visitor, vendor, service, and construction personnel controls
- Alarm systems and response
- Uniformed services duties and responsibilities
- Cameras and other recording systems
- Lock and key control
- Materials control, such as shipping, receiving, and inventory procedures
- Employee security awareness efforts
- Implementation of the drug and alcohol policy according to its specifications
- Emergency planning, such as a review of your evacuation plans
- Hazardous waste materials storage security

CHAPTER 4

Investigation Guidelines

Investigations are conducted to determine whether a company procedure or policy or a civil or criminal law has been violated. Although your company may conduct many different types of investigations, the focus of this chapter will be on security-related investigations.

If you have a corporate security group that oversees your site's security operations, it may be able to help you with investigations in the following three ways:

1. Direction in handling an investigation
2. On-site assistance
3. Guidance in selecting an outside investigator

4.1 GOALS OF AN INVESTIGATION

The goals of an investigation are to

- Gather the facts, determine the truth, and identify the cause of the problem;
- Recover the loss of company assets;
- Analyze the information and identify any trends; and
- Prevent similar losses or occurrences.

4.2 REPORTING TO LAW ENFORCEMENT

When a crime is in progress or when an immediate threat exists to people or property, you may need to contact law enforcement instead of handling the situation solely on your own. Prompt reporting is necessary for the proper and uniform handling of incidents, and will help to

- Protect the rights of individuals;
- Conduct appropriate, accurate investigations;
- Prevent further losses; and

- Lessen the risk of civil liability claims against your company or the individuals directly involved in the matter.

Situations that may require law enforcement notification include theft or damage of employee property, workplace violence, threats of injury, obscene phone calls, violation of company security, organized gambling, substance abuse, and any other criminal act.

Remember that when you contact law enforcement agencies, you must be prepared to cooperate fully with them, and that you may no longer have an internal or administrative solution to an investigation.

4.2.1 Confidentiality

Do not report incidents to anyone who does not have a legitimate need to know. Indiscreet discussion or handling of a report can jeopardize the success of a possible investigation and can also lead to claims of libel, slander, and false arrest.

4.3 DOCUMENT ALL INFORMATION

Information about security incidents should always be documented, no matter how vague or incomplete it might be. Even if the information is so sketchy that it precludes taking any action, it may be useful if additional information is received. This information should be treated confidentially and communicated only to those who have a legitimate need to know.

Management should never ignore or disregard allegations of dishonest or criminal behavior, and should evaluate the source of the information by asking:

- Is the source of the information reliable?
- Is the source in a position to know about the alleged activity?
- Does the source have an ulterior motive for coming forward with this information?
- Does the subject of the allegation have the opportunity and/or the ability to do what has been alleged?
- What is the general reputation of both the source and the subject?

4.4 LEGAL RIGHTS

It is important that every person's rights are protected during an investigation. This applies to suspects, witnesses, and those not involved in any way.

Failure to conduct an investigation in a lawful manner can result in possible claims against the investigator and the company. Investigations should be dealt with quickly and thoroughly. Undue delays may adversely affect innocent employees, hamper the resolution of an incident, and result in the further loss of company assets.

4.5 INVESTIGATIVE RECORDS AND EVIDENCE

This section will help you identify and preserve evidence that may be needed for different types of investigations. You are not expected to be a professional investigator or a forensic expert, but the outcome of an investigation can depend on your knowing what to ask, what to look for, and the types of evidence to preserve.

4.5.1 General Information

Depending on the type of investigation, it may be appropriate to collect the following types of information:

- Personal data
- Name, address, and birth date
- Employee number; hire date; company service, position, division, location, department, supervisor, and performance rating
- Identifying individual characteristics
- Race, sex, height, weight, build, hair color and style, eye color, complexion, clothing, glasses, right- or left handed, or unusual characteristics
- Evidence: Quantity, description, identifying numbers and marks, how the material is marked as evidence, where it is stored
- Stolen or recovered property: Quantity, description, brand, model, identifying numbers and marks, condition, value
- Vehicle license plate number and state; vehicle year, make, model, style, color, value, identifying marks, owner's name
- Weapons/firearms: Type of weapon (knife, club, tire iron, etc.), serial number, make, caliber, type, number of shots, barrel length, model number, color, identifying marks

4.5.2 Physical Evidence

When an incident suggests the presence of physical evidence, your major goal is to preserve it for later professional analysis. When the nature of an incident requires you to protect the scene, follow these general guidelines:

- Avoid disturbing any area that may hold physical evidence.
- Keep spectators away from the area.
- Allow only law enforcement and emergency responders into the area.
- Use a rope or ribbon to mark the area and to deter unauthorized persons from entering.
- Avoid touching anything that may have fingerprints on it. This includes envelopes, letters, and most paper products.

4.6 INVESTIGATIVE RESOURCES

Other departments in your company may be equipped to serve as a resource during your investigation and should always be contacted as soon as possible.

- Auditing assists in reviewing paper trails, authorizations, procedural compliance, records, and financial analysis, when warranted.
- Human resources advises on the employee relations aspects of an incident and possible disciplinary actions to be considered.
- The office of general counsel advises on legal questions.
- Patent counsel handles violations of technical agreements or matters involving intellectual property.
- Insurance determines bond coverage and handles insurance claims.

4.6.1 Outside Assistance

In some cases, it may be appropriate to work with public law enforcement authorities or to hire contract investigative services.

4.7 VOLUNTARY STATEMENTS

Voluntary statements from employees can be an important investigative tool because they help to

- Preserve facts at a time when they are most easily recollected;
- Protect both employees and investigators;

- Provide an accurate, permanent record; and
- Provide for a better follow-up investigation.

Voluntary written statements should be obtained from employees in the case of allegations, admissions, or reports of security-related violations or crimes. These statements may come from people who observed or participated in a violation, or from those whose report is based on rumor, hearsay, or gossip.

4.7.1 Statement Procedure

If a voluntary statement by an employee is considered appropriate, follow these steps:

- Ensure that the statement is handwritten (not typed) by the employee, not by any other individual.
- Instruct the employee to include the time the statement was written (both the start time and finish time) in the upper right corner of the first page of the statement.
- Instruct the employee not to skip lines.
- Instruct the employee to sign, date, and number each individual page in the top margin.
- Have at least one witness—preferably two—sign and date every page at the time the employee signs the statement.
- Provide a copy of the statement to the employee.

4.7.2 Contents of Statement

Dictate the following paragraph to the employee as the first paragraph of the statement:

> I [name] provide the following voluntary statement to [name] that has identified himself/herself to me as [title]. This statement is provided without coercion, or receipt or promise of reward, and is unconditionally submitted.

After this first paragraph is completed, the employee should tell his or her own story, making sure to include all necessary information (who, what, why, when, and how). You may guide the employee to make sure that the proper information is included, but you should not dictate this part of the statement.

At the conclusion of the statement, have the employee write:

> I have read the foregoing consisting of this and [number] other pages. I have initialed each page and all corrections and deletions. This statement is true and correct to the best of my knowledge.

Instruct the employee to write "End of Statement" and sign the document.

4.8 SURVEILLANCE

No employee should ever engage in surveillance outside of normal, routine supervisory responsibilities. Routine supervisory surveillance is limited to monitoring an employee's work performance, such as timeliness, quality, and quantity of work performed.

Investigative surveillance can involve following a person and using cameras or recorders. This type of surveillance requires careful compliance with all state and federal statutes. Investigative surveillance is used only in connection with criminal investigations. Good judgment must precede any decision about the use of investigative surveillance.

4.9 WARRANTS AND SUBPOENAS

Law enforcement personnel and process servers must occasionally enter an individual's place of work in order to serve warrants or subpoenas to him or her. You should always cooperate with law enforcement if they have a search warrant or an arrest warrant. In the case of subpoenas, however, you should contact corporate security (if applicable) or your company's legal counsel first to determine whether you need to make an employee available to the process server.

The security officer or manager should follow these guidelines when it is appropriate to make the employee available:

- Contact the employee and ask the employee to come to a private office or other private, nonwork area. Do not inform the employee of the reason. Avoid taking people into the facility.
- If the employee refuses, notify the employee that your company has a legal responsibility to assist the server. If the employee does not

cooperate, contact his or her supervisor and arrange for the supervisor to escort the server to the employee's work area so the papers can be served.

- If there is concern about the employee's attitude toward being served a warrant or subpoena, escort both the supervisor and the server to the employee's work area.
- The supervisor should arrange to have the papers served in a conference room, office, or other location away from other employees.

4.10 POLYGRAPHS

Laws regulating the use of polygraphs and lie detector devices vary from state to state, are generally very restrictive, and provide severe penalties for illegal tests.

Polygraphs and other tests of this type are generally used only when a loss has occurred and a person suspected of wrongdoing agrees to the test. Many states, however, prohibit such tests even with the consent of the person tested.

In the rare event that there is any possibility that one of these types of tests might be used, always consult your corporate security department (if applicable) or your company's legal counsel.

CHAPTER 5

Inspection Guidelines

The inspection of an employee's or visitor's property can prove tricky for a manager lacking security experience, and could pose liability risks for both the manager and the company if the inspection is performed inappropriately. In this chapter, you'll find general inspection guidelines to consider prior to implementing an inspection policy or conducting an inspection. In addition to these guidelines, the procedure for conducting an interview with the suspected violator of company policy or the law, along with steps for conducting an inspection, is provided. Using the five examples at the end of the "Conducting the Inspection" section, you can hone your skills for recognizing reasonable suspicion to perform an inspection.

In order to understand how inspections should be conducted, it's important to be familiar with the three categories of inspections, including:

1. **Inspections of company property:** These inspections are conducted on a routine basis and are often not security related. It is, however, necessary to understand the difference between company property and the personal property of an employee or visitor.
2. **Inspections based on reasonable suspicion:** These inspections of employees, visitors, and their personal property must be based on the reasonable suspicion of improper conduct by a particular individual.
3. **Inspections based on generalized suspicion:** These inspections of employees, visitors, and their personal property must have the prior approval of your company's management, and are not limited to a specific individual or group of individuals.

Because inspections of company property are routine and not usually security related, the latter two categories of inspections are the focus of this chapter.

In most circumstances, inspections of company property and inspections based on reasonable suspicion will be the only categories of inspections you'll need to conduct as a manager in order to protect company personnel and property. However, sometimes inspections will need to be conducted that are based on a more generalized suspicion that a violation of company policy or the law is occurring or has occurred. Due to the delicate nature of conducting an inspection with a generalized suspicion, you'll find detailed procedures in the second half of this chapter for how to appropriately conduct such an inspection.

5.1 GENERAL INSPECTION GUIDELINES

Important! Because inspections pose many potential civil and criminal liabilities, read these guidelines thoroughly before conducting an inspection.

5.1.1 Designating a Site Coordinator

It's recommended that you designate an inspection policy coordinator (and a backup in the event of his or her absence) who ensures that your inspection guidelines are appropriately implemented. In most cases, this can be the facility manager or security coordinator.

5.1.2 Labor Union Representation

If employees at your company are represented by a labor union, any change in the inspection policy and procedures at the site should be reviewed with the office of general counsel and employee relations before implementation to determine whether and to what extent those changes must be negotiated with the union. It may be necessary to conduct negotiations with the labor union.

5.1.3 Definition of a Visitor

Many of the procedures in this chapter depend on a person's status as a company employee or visitor. For the purpose of these guidelines, the term visitor includes employees of contractors working on company premises, employees of temporary agencies who are working on company premises, and personal and business visitors to company premises.

5.1.4 On-Site and Off-Site Inspections

Inspections should be conducted by company personnel only on property that is owned or leased by your company, or where company work is being performed. If applicable, obtain the prior approval of, and coordinate with, the corporate security department before conducting any inspection at a site that is not owned or leased by your company

5.1.5 Who Should Be Present

Whenever possible, have two supervisors, managers, and/or security personnel present during all inspections. If an employee represented by a labor union requests to have a representative present during any stage of an inspection, the request should be granted. If any employee not represented by a labor union, or a visitor, requests to have a third party present, you're not required to grant the request, but granting the request should be appropriately considered.

5.1.6 Confidentiality

Limit communication and conversations concerning suspicions, the results of inspections, and any refusal by an employee or visitor to cooperate in an inspection to individuals who are on a need-to-know basis.

5.1.7 Suspension and Removal from Premises

In certain circumstances, these inspection guidelines recommend that an employee be suspended or that a visitor be escorted from the premises. If there is evidence that the employee's or visitor's work performance or behavior is impaired, refer to your company's drug and alcohol policy.

If at any time an employee or visitor refuses to cooperate in an inspection conducted in accordance with these guidelines, see the "Refusal to Cooperate in an Inspection" section later in this chapter.

5.1.8 Inspections of Company Property

All supervisory and management personnel have the right to inspect company property at any time for any reason falling within the normal scope of their authority and job responsibilities, with one exception: an area of company property where employees have an expectation of privacy (described in the following section).

Note: Because inspections of company property are conducted on a routine basis, this playbook does not recommend procedures to be followed in conducting such inspections.

5.1.9 Expectation of Privacy

There is one exception to searching company property: you are not allowed to search company property that is designated for use by employees or has been permitted for use by employees in a manner that provides employees with some expectation of privacy.

For example, unless appropriate notice is given to employees, they will probably have a reasonable expectation of privacy with respect to the contents of their lockers. For this reason, an employee's locker should be treated in a manner similar to personal property of the employee, and should only be inspected by following the guidelines for one of the other two types of inspections: those based on reasonable or generalized suspicion. The same is true for any other company property where employees have a reasonable expectation of privacy.

5.1.10 Is It Company Property?

On occasion, it may not be clear whether certain property is company property or the personal property of an employee. For example, some employees may use briefcases for which they have been reimbursed by your company. Where any doubt exists, the property should be regarded as personal property of the employee.

5.1.11 Reasonable Suspicion

All supervisory, management, and security personnel have the right to conduct inspections of employees, visitors, and their personal property based upon reasonable, individualized suspicion of improper conduct. The following guideline describes what constitutes reasonable suspicion and the procedures for conducting inspections based on reasonable suspicion.

5.1.12 What Is Reasonable Suspicion?

Reasonable suspicion exists when specific objective facts and rational inferences drawn from those facts in the light of experience lead to a suspicion that a particular person has violated or is violating company policy or the law.

Information concerning a possible violation may come to a supervisor or manager either through direct observation or from a third party. In either case, the supervisor or manager must evaluate that information to determine what rational inferences can be drawn from it based upon the supervisor's or manager's experience.

Where information comes from a third party, the supervisor or manager must also evaluate the credibility of the information and the reliability of the third party, and must attempt to obtain verification of information obtained.

Later in the chapter, examples are provided to show you how reasonable suspicion may be determined.

5.1.13 When an Inspection May Be Conducted

Inspections based on reasonable suspicion may be conducted when there is either:

- A reasonable suspicion that an employee or visitor is in violation of company policy, or
- A reasonable suspicion that an employee or visitor is in violation of the law and such violation presents a risk to company safety, health, productivity, property and/or security, public safety, health or property, or the safety, health, or property of any company employee.

5.1.14 Is an Inspection Appropriate?

If an employee's or visitor's behavior, performance, or involvement in one or more accidents leads a supervisor or manager to suspect that the employee or visitor may be in violation of company policy or the law, the supervisor or manager should perform the following:

- Record the behavior, performance, accident(s), or other data, and the names of any witnesses.
- If any of the information that led to the suspicion came from a third party, attempt to verify the information.
- Consult with another supervisor or manager (when an employee is under suspicion this should be the employee's direct supervisor if available and not already involved) to confirm that a reasonable suspicion exists that a violation of company policy or the law has occurred or is occurring, and that an inspection is appropriate. This consulting supervisor or manager may also serve as a witness or participant if any inspections are conducted.

5.1.15 Limits of Search

Conduct inspections only where there is reasonable suspicion that a violation will be found. For example, having a reasonable suspicion that an employee has placed stolen company property in his or her locker does not necessarily constitute reasonable suspicion that stolen company property will be found in the employee's vehicle.

5.2 INSPECTIONS BASED ON REASONABLE SUSPICION

The following procedures for interviewing the suspected violators of company policy or the law and performing an inspection relate specifically to inspections based on *reasonable* suspicion. (Inspections based on *generalized* suspicion are discussed separately later in this chapter.)

5.2.1 Interview Procedures

After reviewing the general inspection guidelines and determining that an inspection based on reasonable suspicion is appropriate, in conjunction with the supervisor or manager of the suspected violator you should:

- In a quiet, nonintrusive manner designed to avoid drawing the attention of others to the suspected employee or visitor, ask the employee or visitor to accompany you to a private office (or, if an office is not available, to a safe, private, nonwork area) for an interview.
- Inform the employee or visitor that you suspect that the employee or visitor may be in violation of or may have violated company policy or the law.
- Inform the employee or visitor that he or she is not being accused of wrongdoing, but you have received information that your role as a supervisor or manager requires that you investigate. Be very careful not to accuse the employee or visitor of wrongdoing or to state or imply that you've already determined the employee or visitor is guilty; for example, you shouldn't say that you know or have proof that the employee or visitor has violated or is violating company policy or the law.
- Describe the suspected violation. Inform the employee or visitor of the basis for the suspicion. If questioning is appropriate, limit questioning to the facts and/or circumstances that led to a suspicion that a violation of the law or company policy has occurred. Give the employee or visitor an opportunity and a reasonable period of time to answer any questions, offer an explanation, or make any comments.
- If the interview lasts for an extended period of time, lunch or other appropriate breaks should be allowed.

5.2.2 Conducting the Inspection

If you still believe that an inspection is appropriate following the interview(s), inform the employee or visitor that you would like to conduct an inspection, describing the scope of the inspection you intend to conduct (for example, clothing, locker, briefcase, and/or vehicle). Answer any reasonable questions concerning the nature and extent of the inspection.

If it is determined that an inspection of the employee's or visitor's clothing is appropriate, the inspection should be conducted in the following manner:

- The inspection should be conducted by a person of the same sex as the employee or visitor. No member of the opposite sex should be present. If possible, at least one witness should be present.
- The inspection should be limited to a request that the employee or visitor empty his or her pockets, surrender for inspection any handheld or non-clothing item carried on his or her person, and, if appropriate, remove shoes and/or stockings. No pat-down or other inspections involving physical contact should be conducted.
- If it is determined to be appropriate, inspect the employee's or visitor's handheld and other non-clothing items, locker, and/or vehicle.
- No action should be taken intentionally that results in damage to vehicles, briefcases, lunch boxes, and other personal property of the employee or visitor. If the employee or visitor refuses to permit inspection of a locked item of personal property, do not use force—follow the guidelines in "Refusal to Cooperate in an Inspection," discussed later in this chapter. The only exception to this rule is when an inspection of a locker is determined to be appropriate: a lock owned by the employee or visitor may be cut off a locker provided your company pays for a replacement lock.
- If the inspection reveals what appears to be a prohibited material, or what appears to be company property in the employee's or visitor's possession or control without apparent authority or permission, or some other apparent violation of company policy or the law, follow the guidelines listed in the "Discovery of a Suspected Violation" section later in this chapter.
- If the inspection does not reveal any apparent violation of company policy or the law, return the employee to work or the visitor to an appropriate location and assure the individual that the inspection is over.

- If applicable, inform corporate security of the incident if they have not already been informed.
- Prepare a detailed written report of the incident.
- Limit conversations about the incident to a need-to-know basis.

5.2.3 Disciplinary Action Following the Inspection

Disciplinary action, if appropriate, may be taken when a policy violation is discovered, **even when the violation discovered is not the suspected violation that prompted the inspection.** For example, if you have reasonable suspicion that an employee is in possession of alcohol in violation of company policy, disciplinary action may be taken if an appropriate inspection instead reveals that firearms have been brought onto company property.

5.2.4 Determining Reasonable Suspicion

Use the following five examples to test your knowledge of how to determine what constitutes reasonable suspicion.

Example 1: A detailed, believable report from a third party indicates that an employee is violating company policy. The third party has a history and a reputation for truthfulness, and is an independent observer with no stake in the outcome of any actions that may result from the information provided. The third party has no apparent reason under the circumstances to exaggerate or falsify the information provided.

Q Should the supervisor or manager attempt to obtain verification of the information?

A Yes. Reasonable attempts should be made to verify any information received from a third party. Verification may come from either direct observation or from other reliable third parties.

Q If verification is not obtained, does reasonable suspicion exist?

A Under the circumstances described above, assuming that no contradictory information was discovered in the supervisor's or manager's attempt to obtain verification, reasonable suspicion probably exists even though it is based on a single report from a third party.

Example 2: An anonymous telephone tip indicates that an employee is violating company policy. A similar report is received from an employee who is known to have a grudge against the accused employee.

Q Does reasonable suspicion exist that the accused employee is violating company policy?

A No. Each of these reports should be considered to be of almost no reliability. Two highly unreliable reports cannot serve as verification of each other.

Q Could these reports lead to a reasonable suspicion of a policy violation?

A Yes, if verified by direct observation by a supervisor or manager or if verified by believable information obtained from reliable third parties, and if rational inferences drawn from the verified information lead to a suspicion that the identified employee is violating company policy.

Example 3: A supervisor enters a bathroom and finds what he or she believes to be cocaine on a shelf. The supervisor has been trained to recognize cocaine. One employee is currently in the bathroom but is not near the substance that appears to be cocaine.

Q Does reasonable suspicion of a policy violation by this employee exist based solely on this information?

A Under the limited facts given in this example, no. Even if the substance on the shelf is tested and found to be cocaine, there is not enough information in this example to result in a reasonable suspicion that the employee found in the bathroom at the time of its discovery has been in possession of cocaine on company property. The cocaine could have been left by a previous occupant of the bathroom.

Example 4: Reliable, verifiable information leads to a rational inference that two people within a group of five have engaged in a sale of illegal drugs on company property.

Q Does reasonable suspicion of a policy violation exist to justify a search of all five employees?

A No. In order to have reasonable suspicion that all five employees were involved, reliable information must indicate each individual person suspected of violating company policy or the law.

Example 5: A supervisor observes an employee placing what appears to be company property into the employee's car. The property is not of a type that employees normally can be expected to have permission to remove from the work site.

Q Should the supervisor attempt to obtain verification of his or her observations?

A Verification is not required where a supervisor directly observes information that clearly indicates a possible policy violation. If the supervisor has doubts concerning the reliability of his or her observations, verification may be appropriate.

Q Does reasonable suspicion of a policy violation exist?

A In this case, reasonable suspicion of a policy violation probably exists. In all cases where a supervisor or manager intends to take action based on reasonable suspicion, the supervisor or manager should, if possible, consult with another supervisor or manager to obtain a second opinion to confirm reasonable suspicion exists. This consulting supervisor or manager can also participate in or witness any action taken by the supervisor or manager.

5.3 INSPECTIONS BASED ON GENERALIZED SUSPICION

Where some risk or generalized suspicion of improper conduct exists, it may be desirable to conduct an inspection that does not fall under either of the first two categories of inspection named in the introduction to this chapter (inspections of company property and those based on reasonable suspicion). Conducting this type of inspection should be conducted only with the approval of your company's management. This section describes the circumstances where this type of inspection may be appropriate.

5.3.1 Reasons for Inspection

In collaboration with the inspection policy coordinator and other appropriate company management, you should identify the type(s) of inspection desired and the reasons why the inspection is needed to protect company property or the safety and health of employees or the public.

Factors that may be considered when planning an inspection based on generalized suspicion include

- Past problems such as excessive absenteeism, high accident rates, poor productivity, erratic work quality, theft, and other safety, health, productivity, and/or security problems;
- Current risks to safety, health, property, and/or security;

- The degree of intrusiveness of the type of inspection under consideration; and
- The expected degree of effectiveness of the type of inspection under consideration.

If it is determined that an inspection is appropriate based on a generalized suspicion, the basis for that determination should be documented and submitted for review and approval by human resources, employee relations, the office of general counsel, the operations manager responsible for your facility, and corporate security (if applicable).

If approval as described above is obtained for inspections based upon generalized suspicion, provide written notification to employees 30 days before beginning the inspections.

5.3.2 Types of Inspections

After you have determined that an inspection based on a generalized suspicion is necessary, you have obtained approval from the appropriate departments and parties in your company, and you have provided a 30-day written notice to employees, you may conduct the following three types of inspections:

1. Handheld items: The contents of lunch boxes, briefcases, handbags, toolboxes, packages, and similar items, but only upon entrance to and/or exit from buildings located on company premises.
2. Lockers: The contents of lockers on company property.
3. Vehicles: The contents of vehicles, but only upon entrance to and/or exit from company premises.

5.3.3 Inspection Type #1: Handheld Items

Post the following notice, or a substantially equivalent one, at each entrance to any building where the entrance and/or exit inspections will be conducted. Make sure the notice is large enough to be clearly visible and easily read.

> **NOTICE**
>
> To help assure safety at this site and to protect company property, designated company personnel may inspect the contents of lunch boxes, briefcases, handbags, toolboxes, and other packages and containers entering [or exiting] this building. These inspections may occur at any time without prior

announcement. YOUR ENTRANCE INTO THIS BUILDING CONSTITUTES YOUR CONSENT TO THESE INSPECTIONS. Failure of an employee to permit inspection may result in disciplinary action, including termination of employment, even for a first offense. If any visitor refuses to permit inspection, that person may be barred from this site, and other appropriate action may be taken.

5.3.4 Inspection Type #2: Lockers

Post the following notice, or a substantially equivalent one, immediately above, adjacent to, or at the entrance to the locker area. Make sure the notice is large enough to be clearly visible and easily read.

NOTICE

To help assure safety at this site and to protect company property, designated company personnel may inspect the contents of these lockers at any time without prior announcement. YOUR USE OF ANY OF THESE LOCKERS CONSTITUTES YOUR CONSENT TO THESE INSPECTIONS. Failure of an employee to permit inspection may result in disciplinary action, including termination of employment, even for a first offense. If any visitor refuses to permit inspection, that person may be barred from this site, and other appropriate action may be taken.

5.3.5 Inspection Type #3: Vehicles

Post the following notice, or a substantially equivalent one, at each entrance to any parking area, lot, or other area where the entrance and/or exit inspections will be conducted. Make the notice large enough to be clearly visible and easily read.

NOTICE

To help assure safety at this site and to protect company property, designated company personnel may inspect the contents of vehicles entering [or exiting] this area. These inspections may occur at any time without prior announcement. YOUR ENTRANCE INTO THIS AREA CONSTITUTES YOUR CONSENT TO THESE INSPECTIONS. Failure of an employee to permit inspection may result in disciplinary action, including termination of employment, even for a first offense. If any visitor refuses to permit inspection, that person may be barred from this site, and other appropriate action may be taken.

5.3.6 Inspection Pattern

Limit the discretion of those conducting the inspections. For example, inspect the packages, lockers, or vehicles of all persons; all persons entering or exiting during a particular time period; all persons using a particular entrance/exit; or every tenth person, etc. The pattern of inspection does not need to be the same each time, but inspectors should not have discretion to pick and choose which persons' packages, lockers, or vehicles are inspected.

If an inspection reveals the presence of what appears to be a prohibited material, or if it appears that company property is being removed from the site without apparent authority or permission, or if some other apparent violation of company policy or the law is revealed, follow the procedures outlined in "Discovery of a Suspected Violation," discussed below.

5.3.7 Discovery of a Suspected Violation

Most violations of company policy or law that are discovered during inspections fall into three categories:

1. Alcohol and/or drugs
2. Theft
3. Firearms

Upon discovery of what appears to be a violation of company policy or the law, be very careful not to accuse the employee or visitor of wrongdoing, or state or imply that you have already determined that the employee or visitor is guilty. For example, never say to the suspected violator that you know or have proof that the employee or visitor has violated or is violating company policy or the law.

Evidence of the sale, transfer, purchase, or possession of illegal drugs; theft; or other illegal acts may be referred to local law enforcement officials if you and your inspection team determine that such action is appropriate in light of the facts and circumstances.

Law enforcement authorities may also be involved in situations where an employee or visitor poses a threat to himself or herself, to others, or to company property or equipment. Events that constitute violations of company policy but are not violations of the law, and do not create a suspicion that a violation of the law has occurred, should not be reported to law enforcement officials.

In the following sections of this chapter, guidelines for dealing with each of the three categories of company policy or law violations are provided. Whenever possible, two supervisors, managers, and/or security personnel should be present during all actions taken under the specific procedures below.

5.3.8 Violation Category #1: Alcohol and Drugs

5.3.8.1 Apparently Legal Drugs

When inspection reveals the presence of what appears to be a *legal* prescription or over-the-counter medication, record the name of the drug and the quantity. Depending on the medication and the circumstances, it may be appropriate to request proof of a prescription, particularly if an employee's job function requires that the use of prescription medication be reported to the company's medical services/human resources representative, and the medication has not been reported.

Return the medication to the employee or visitor unless there is reasonable suspicion that the substance is an illegal drug, even though it appears to be a legal drug.

If reasonable suspicion exists that the substance is an illegal drug, take these steps:

- Remove a single sample (or a small quantity in the case of liquids, powders, etc.) of the suspected material for testing.
- Place the sample in a sealed container marked for identification.
- Record the quantity of the remaining material.
- Return the remaining material to the employee or visitor.
- Complete the inspection. (See "Test Results" later in this section for further guidelines.)

5.3.8.2 Alcohol and Illegal Drugs

If an inspection reveals what appears to be an *illegal* drug or alcohol in violation of company policy, seize the suspected material. If not in a private area, return with the employee or visitor to a private office or a safe, private nonwork area to interview the employee or visitor regarding the nature of the suspected material. The employee or visitor should be given an opportunity and a reasonable period of time to offer any explanation.

If an employee or visitor identifies the substance as alcohol (and possession in the manner discovered is in violation of company policy) or an illegal drug, place the seized material in a sealed container marked for identification and record the quantity of the material and the identity given by the employee or visitor. If the material is identified by the employee or visitor as a prescription medication, it may be appropriate to request proof of prescription as described earlier under "Apparently Legal Drugs."

After you've found an employee or visitor to be in possession of alcohol or illegal drugs, except when you have determined that no violation of company policy or law has occurred, take the following steps:

- In the case of an employee, suspend the employee pending a determination by the appropriate management committee as to whether and to what extent discipline should be imposed.
- In the case of a visitor, record the visitor's name and other information (such as employer, company employee being visited, etc.) appropriate for identifying the visitor and escort the person from the premises.
- Contact your medical services department (if applicable) or local authorities to arrange for laboratory testing to identify any suspected substance(s) found during the inspection.

5.3.8.3 Test Results

If test results show that a substance seized during an inspection is an illegal drug, the substance should be turned over to an appropriate law enforcement agency even if other details of the incident are not reported.

In the case of an employee, meet with the employee as soon as test results are available and inform the employee of the test results. If the results are negative, return the employee to work. Provide back pay if appropriate. If appropriate, assure the individual that the investigation is over. If test results reveal a prohibited substance, continue the suspension until the appropriate management committee determines whether and to what extent disciplinary action is appropriate.

In the case of a visitor, unless the visitor cannot be located, inform the visitor of the test results as soon as test results are available. If the results are negative, inform the visitor that the investigation is over, if appropriate. If test results reveal a prohibited substance, take the following steps:

- Inform the visitor that he or she is barred from entering company premises in the future.

- If at the time of the inspection the visitor was on company premises in the conduct of business for the visitor's employer, inform the visitor's direct supervisor of the test results, the fact that the visitor is barred from company premises, any other action taken by your company that would affect that employer, and the reason for your actions.

Finally, always prepare a detailed written report of the incident. Limit conversations concerning the incident to a need-to-know basis.

5.3.9 Violation Category #2: Theft

If inspection reveals that the employee or visitor is in the possession or control of what appears to be company property without apparent authority or permission (such as a materials pass), do the following:

- Seize the material, store it in a secure location, record the quantity and description of the material, record the location where the seized material is stored, record the identity of the person from whom the material was seized, and indicate to the employee or visitor that you are seizing the material for investigation. If the material cannot be seized or stored, secure the material in place and take photographs of the location of the material.
- Return with the employee or visitor to a private office or safe, private nonwork area. Interview the person regarding the nature of the apparent company property and the reason for its possession or control by the employee or visitor. The person should be given an opportunity and a reasonable period of time to offer any explanation. Investigate any explanation given concerning the property, except when further investigation is unnecessary because the person admits theft.
- Complete the inspection.

If an employee admits theft, suspend the employee pending a determination by the appropriate management committee of appropriate disciplinary action. If a visitor admits theft, do the following:

- Record the visitor's name and other information (such as employer, company employee being visited, etc.) appropriate for identifying the visitor.
  - Inform the visitor that he or she is barred from entering company premises in the future.
  - Escort the visitor from the premises.

- If at the time of the inspection the visitor was on company premises in the conduct of business for the visitor's employer, inform the visitor's direct supervisor that the visitor is barred from company premises. Also inform the employer of any other action taken by your company that would affect that employer, and the reason for the company's actions.
- As with other allegations of the violation company policy or the law, prepare a detailed written report of the incident and limit conversations concerning the incident to a need-to-know basis.

5.3.10 Violation Category #3: Firearms

If an inspection reveals the presence of firearms in the possession or control of an employee or visitor, do the following:

- A person trained in the use of firearms should inspect the firearm for presence of ammunition. Remove any ammunition present and any other ammunition revealed by the inspection. Place the ammunition in a sealed container marked for identification, and place the container in a locked safe, drawer, or office.
- Place the firearm in a different sealed container, marked for identification if feasible, and place the firearm in a locked safe, drawer, or office separate from the ammunition. Record the location of the ammunition and firearm and the identity of the person from whom they were seized.
- Complete the inspection.
- In the case of an employee, suspend the employee pending a determination by the appropriate management committee of appropriate disciplinary action. Escort the employee from the premises.
- In the case of a visitor:
  - Record the visitor's name and other information (such as employer, company employee being visited, etc.) appropriate for identifying the visitor.
  - Inform the visitor that he or she is barred from entering company premises in the future.
  - Escort the visitor from the premises.
- If at the time of the inspection the visitor was on company premises in the conduct of business for the visitor's employer, inform the visitor's direct supervisor that the visitor is barred from company premises. Also inform the employer of any other action taken by your

company that would affect that employer, and the reason for the company's actions.

- Deliver the firearm and ammunition to local law enforcement officials for retention or return them to the employee or visitor.
- Prepare a detailed written report of the incident, and limit conversations concerning the incident to a need-to-know basis.

5.4 REFUSAL TO COOPERATE IN AN INSPECTION

The following guidelines are intended to assist supervisors and managers in dealing with an employee or visitor who refuses to cooperate in an inspection.

5.4.1 General Guidelines

Deception, force, and other improper means should not be used to obtain cooperation from any individual who refuses to cooperate. Force may be used only to the extent necessary for self-defense, the defense of others, or the protection of company property. If the use of force is or appears to be required, contact other site personnel who assume security responsibilities and/or local law enforcement officials as soon as possible to assume control of the situation.

5.4.2 Guidelines for Employee Refusals

- If an employee refuses to cooperate in an inspection, place the employee on suspension without pay for an indefinite period pending review by appropriate local management.
- A detailed written report of the incident should be prepared by those involved.
- Management should promptly (within three working days, if possible) determine appropriate disciplinary action.

5.4.3 Guidelines for Visitor Refusals

- Record the visitor's name and other information (such as employer, company employee being visited, etc.) appropriate for identifying the visitor.
- Inform the visitor that he or she is barred from entering company premises in the future.
- Escort the visitor from the premises.

- If at the time of the inspection the visitor was on company premises in the conduct of business for the visitor's employer, inform the visitor's direct supervisor that the visitor is barred from company premises. Also inform the employer of any other action taken by your company that would affect that employer, and the reason for the company's actions.

CHAPTER 6

Emergency Procedures

The effectiveness of your actions in an emergency depends on how well you have developed emergency plans. Your plans should cover three phases:

1. Responding immediately
2. Controlling the situation and finding a remedy for it
3. Returning to normal business operations

6.1 TYPES OF EMERGENCIES

An emergency is any situation that could result in an immediate safety or health threat to a person, or in damage to property or the environment. This situation can result from a variety of accidental or intentional causes:

- Criminal and security-related incidents that escalate to a local emergency, such as bomb threats, arson, sabotage, civil strife, or property destruction
- Serious injury or illness
- Property damage from fire or explosion
- Environmental discharge accidents, such as chemical spills or leaks
- Natural disasters, such as floods, tornadoes, earthquakes, and winter storms
- Conditions that require a shutdown of operations, such as a roof collapse
- Incidents of potentially high community interest

6.2 SECURITY CONCERNS

Safety is a primary concern in emergencies, but these situations also raise immediate security issues, including the following:

- Appropriate staffing of security officers
- Alarm service and emergency response
- Site and scene protection

- Property and information protection
- Law enforcement contact and coordination
- Emergency vehicle access
- Crowd control
- Special access controls for extended situations

6.3 MANAGING WITH A PLAN

A formal plan can help you manage and control emergency situations. Much of your plan may focus on safety issues, but there will also be some important security components (which your corporate security department could help you with, if applicable). Emergency planning is a seven-step process, as described below.

6.3.1 Step 1: Examine Your Vulnerability

Examine the likelihood of various emergencies and your vulnerability to them. Identify the situations that could have a sudden and negative impact on your operations.

6.3.2 Step 2: Prepare Your Plans

Plan for the emergencies that are most likely to occur.

If possible, reduce the probability that the emergencies will occur. Determine who will make what decisions under what circumstances. Establish an emergency center or other location where information will be channeled and decisions made.

Train an emergency response team appropriate to each emergency:

- Spill team
- First responders
- CPR units
- Computer and data recovery team
- Written records recovery team
- Boiler recovery team

Identify alternate means of communication, and define the conditions that will require evacuation or relocation of your operations.

6.3.3 Step 3: Ensure Recovery of Operations

An organized recovery of operations depends on continuity of management and critical functions. Identify your primary and alternate sources for critical materials and spare parts. Plan for the preservation and recovery of essential records.

6.3.4 Step 4: Coordinate the Plan

Discuss the plans with all concerned groups in your facility, and with any outside agencies you may need to call on to help verify your assumptions. If your plan depends on police and fire support, contact them to make sure of their cooperation.

6.3.5 Step 5: Test the Plan

Testing your plan will help you sort out any problems and will also prepare your employees for a real emergency. You can test your plan in stages or through a full participation exercise.

6.3.6 Step 6: Implement the Plan

In an actual emergency, stick to your plans as closely as possible. It is especially important that decision makers get timely and accurate information in order to assess the situation. Contain the situation as quickly as possible, and begin recovery operations as soon as the emergency is under control.

6.3.7 Step 7: Evaluate the Plan

Evaluate your plans thoroughly and objectively after each test. After a real emergency, perform an evaluation as soon as possible after you resume normal operations.

6.4 REPORTING

When an emergency arises, notify your company's management team as soon as possible. After the emergency has been handled, you should write a formal written report. After completing the report, be sure to forward it to management for their records. Some of the forms you also may need to complete are: injury reports, fatality reports, and Environmental Protection Agency (EPA) reports.

6.5 EVACUATION

Emergencies may sometimes require an evacuation of your facility. It's important that you identify the circumstances that will require an evacuation, and that you establish and practice evacuation procedures.

Specific evacuation procedures may differ in various emergencies, but your planning should consider these general guidelines:

- Determine who can order an evacuation. Public authorities such as police, fire, health, civil, and EPA officials always have this authority, but you should determine which members of your management may declare an evacuation.
- Determine in advance what circumstances or conditions would require an immediate evacuation.
- Determine the evacuation signals and other means of communications you will use.
- Plan evacuation routes carefully so that employees are not directed through or near hazardous areas.
- Prominently display evacuation routes and instructions throughout your facility. Mark emergency exits clearly.
- Post warnings on elevators advising that they should not be used during an evacuation.
- Designate employees to monitor specific posts in the facility, guide employees, and make final sweeps to ensure that their areas are clear. Select people who are generally available, and who can provide calm, authoritative leadership. Designate alternates, and update the monitor rosters frequently.
- Establish waiting areas outside the facility. This will help you account for all people and keep employees at a safe distance from the facility and out of the way of emergency equipment and personnel. Designated waiting areas are also useful when you need to communicate further instructions to the employees.
- Do not allow employees to re-enter the site for any reason until management or the public authorities determine it is safe.

6.6 BOMB THREATS

Awareness and planning can do much to reduce the tension and confusion that accompany a bomb threat. According to the Federal Bureau of Investigation (FBI), in 95 percent of bomb threats no device is ever found. On the other hand, most actual bombing incidents are not preceded by any type of warning. Despite these statistics, you should make plans to deal with bomb threats on a case-by-case basis.

6.6.1 Planning

An important part of planning is to contact your local police department (and your local military resources if that is recommended or if no local police are available). Determine the conditions under which your local police or military resources will help you search the facility. They may be able to provide you with dogs trained in such searches, as well as expert help in identifying and removing devices.

6.6.2 Phone Threats

Because most bomb threats are received by phone, the people who handle your incoming calls should be instructed to do the following:

- Remain calm.
- Ask what particular facility is being threatened.
- Ask what time the device is set for.
- Ask specifically where the bomb is located.
- Ask why the caller is taking this action.
- Record any background noise, unusual voice characteristics, or speech mannerisms.
- Record the date and time of the call and which phone number it came in on.
- Tell only designated managers about the call.

6.6.3 Assess Credibility

When there is a bomb threat, the first thing you need to do is assess the threat's credibility and determine whether further action is needed. Calls that are short and nonspecific generally do not have sufficient credibility to merit an extraordinary response on your part, but you should always document the incident and report it to management or your corporate security department.

Calls that are long, contain specific about the type of device, and touch on specific grievances deserve further consideration.

You should also consider whether the caller could have gained access to place the device, and if conditions in the facility or the surrounding community are likely to produce an actual bombing.

If you determine the threat is credible, you may need to consider several additional factors. Do you have the time and resources to conduct a search first? Does the nature of your business include any critical operations (handling hazardous chemicals, for example) that would require immediate evacuation? The credibility of the threat and your assessment of these issues will help you determine whether evacuation is necessary.

In addition to whatever assistance your local law enforcement and military resources can provide, you may also have to form volunteer employee search teams. Volunteers should be assigned to search areas they are familiar with, beginning with the most accessible areas and ending with the least accessible.

If a suspicious device is found, search team members must not touch it, but should contact management and report its location and appearance. They should then shut down any machinery in the area and depart immediately. Notify the authorities if they are not already on the scene. Do not let anyone reenter the building until authorities consider it safe.

For a visual summary of the information in this section, see the model for responding to bomb threats (Figure 6.1).

6.6.4 Bomb Threat Search Diagrams

In Figure 6.2, you'll see that bomb searches should be conducted according to specific procedures to ensure thorough and accurate coverage of all locations in your company's building(s). The top diagram illustrates the pattern with which each room should be swept: search teams should be broken up, and then cover the room in two groups. The bottom diagram illustrates the three levels of search that should be performed in each room:

- Level One: the area from the floor to the searcher's waist
- Level Two: the area starting at the searcher's waist and ending at the top of the searcher's head
- Level Three: the area from the top of the searcher's head to the ceiling

6.7 CIVIL DISTURBANCES AND DEMONSTRATIONS

Frequently viewed as a product of 1960s and 1970s social movements, demonstrations continue to be a means of expression about any number of valid issues, such as abortion, apartheid, the environment, and nuclear weapons. The primary objective of demonstrators is usually media

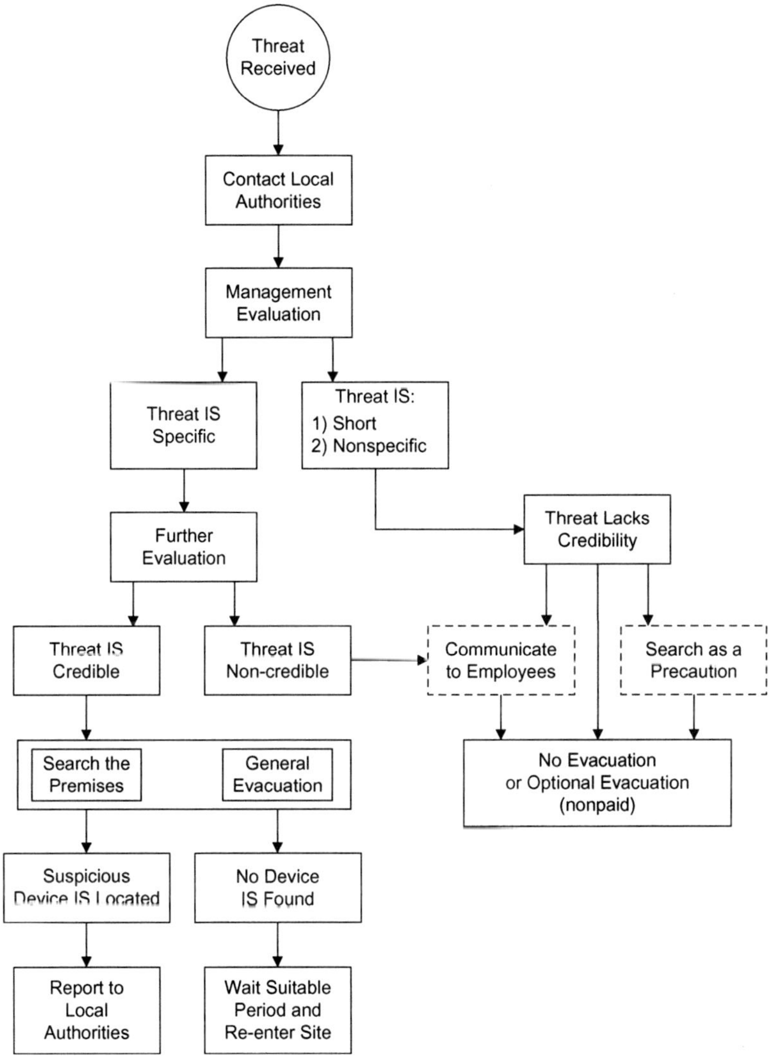

Figure 6.1 This bomb threat response model will help you follow the specific procedures for responding to bomb threats called in to your company, including: notification of appropriate parties (law enforcement, management, employees); determining whether the threat is credible; ordering employee evacuations; and searching company property for the bomb device.

attention; communicating with the company is often secondary. Related objectives may include getting arrested by trespassing on company property, engaging the company in public debate, conveying a message directly to employees, or, occasionally, inciting physical confrontation.

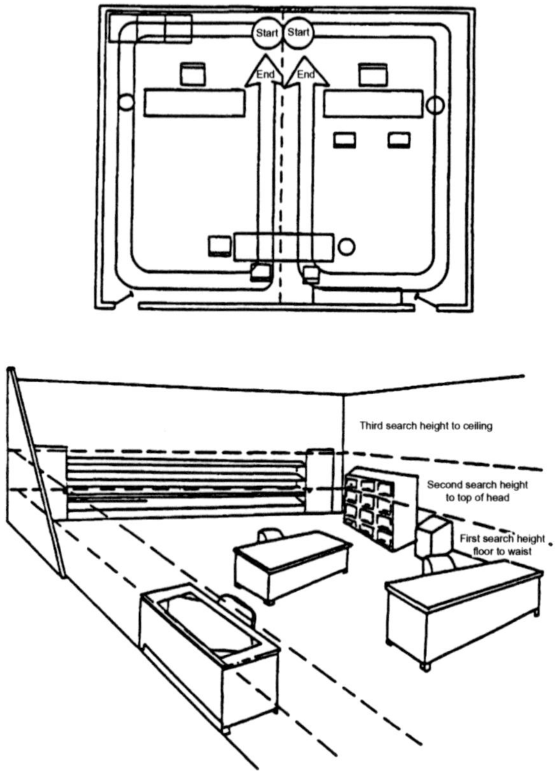

Figure 6.2 The two diagrams in Figure 6.2 demonstrate the procedure for proper bomb searches. Rooms should be divided among the search volunteers (top diagram), and should be searched beginning with the floor and ending with the ceiling (bottom diagram).

Every organization should prepare and implement a demonstration plan (to be incorporated into the overall security plan) for handling demonstrations on or near company property. As the person responsible for the safety of company employees and assets, you should make sure to communicate in your demonstration plan that company grounds are private property maintained for the use of company employees, customers, vendors, and guests. Use of company premises for demonstrations interferes with the normal conduct of business and is, therefore, not allowed.

The best defense in the case of demonstrations is to be aware of conditions in your community and make sure you are notified of any actions that could affect operations. Your public relations department and other governmental and business sources can also keep you informed.

6.7.1 Building a Demonstration Plan

The objectives of your plan should be to

- Avoid injuries to individuals and damage to property;
- Prevent harassment of employees and interruption of work;
- Minimize the possibility of confrontation and arrest;
- Correctly state your company's position on the issue;
- Maintain effective relations with the news media and public officials; and
- Avoid potential legal problems.

As you build a demonstration plan, remember to perform the following nine steps:

1. Review company policy and applicable local ordinances and laws with local law enforcement officials.
2. Confirm the boundary lines between company and public property.
3. Negotiate with local law enforcement officials for an acceptable public area to allow demonstrations. Try to select an area near a vantage point, such as a window, from which you can photograph the event but not be observed.
4. You might consider purchasing video cameras and recorders to document the entire event. At a minimum, have a camera available.
5. Evaluate perimeter and access controls to the facility, and modify procedures if additional access control is needed.
6. Designate a security coordinator or other manager to be the initial contact with the demonstration leaders.
7. Select a room where company representatives and the demonstration leaders could meet if necessary, and a room for news media briefings.
8. Establish a notification list. This should include anyone who may need to know about potential or actual demonstrations (i.e., business unit managers, company management, or local officials).
9. Make sure that your colleagues involved with security operations are briefed on how to handle demonstrators.

6.7.2 Task Force Responsibilities

If your company is a part of a larger corporation, demonstrations or protests will require a coordinated response by facility management (you), division or subsidiary management, and corporate staff. This coordination is the responsibility of a task force that consists of the persons in charge of administration, corporate quality and manufacturing services, and public relations. The task force determines policy and provides guidance and any corporate assistance to the facility. Selected representatives of affected staff departments or divisions will advise the task force, implement its policy decision, and act in place of the task force if its members are not immediately available.

6.7.3 Assess Impact

When you hear about a potential disturbance, you must assess its probable impact on operations. Consider these factors:

- How serious or widespread will the disturbance be?
- Will employee access to the facility be hampered? Can you arrange alternate routes and modes of transportation?
- Are perimeter safeguards adequate to protect the facility? What steps can you take to protect vulnerable areas?
- Will supplemental security patrols be needed?

6.7.4 Reporting

Immediately report any warning or actual organizing or gathering of a demonstration to your local management and/or your corporate security department. These groups will notify the appropriate departments, including the office of general counsel. These departments might also send observers or other assistance to the scene if necessary.

6.7.5 Local Management Response and Control

As the company representative at the scene, you should take charge of the situation and be the main contact with demonstrators, the police, and other company groups. Specifically, you should adhere to the following guidelines for responding to a demonstration:

- If applicable, keep corporate security informed.
- Keep local police advised of the situation. Request their assistance to monitor or control the situation if necessary.
- Keep employees informed. Recommend they avoid contact or confrontation with demonstrators.
- Follow the established publicity procedures for emergency situations. If you are contacted by reporters, ask them who they represent and exactly what they need to know, and tell them that you or another qualified person will call them back. Then contact a public relations representative about handling the reply.

- Do not allow the demonstrators to use any company facilities, either buildings or parking.
- No company employee or security manager should physically contact or try to arrest demonstrators. Leave arrests to local law enforcement officers.
- The security coordinator or designated company representative should make sure the event is being videotaped or photographed to supplement written reports. The person with the camera should record the event from a vantage point where it's possible to maintain a low profile.
- Consult with the demonstration task force or designated representatives as appropriate.

Your company's public relations representative or group is responsible for contact with the news media and community officials, and for communications with the demonstrators regarding the company's position on the issue.

Public relations is also responsible for advising and assisting facility management on communications with the news media, employees, demonstrators, and community officials, or sending a representative or arranging public relations assistance for local management as appropriate.

Public relations should also prepare any official company statements to the news media and employees.

6.7.6 Debriefing

In conjunction with other managers and staff departments, always review and evaluate your actions and response with the task force once the demonstration has ended.

6.8 NATURAL DISASTERS

6.8.1 Communications

Communications can be a critical factor in any natural disaster. It's important to be prepared for an interruption of normal lines of communication. For example, if land phone lines are down, you can try using a wireless phone. If neither of those are working, sometimes text messaging or email will still work. In recent years, communication through the use of social media (i.e., Facebook or Twitter) has been a critical way that individuals and companies affected by a natural disaster have communicated with the outside world and solicited assistance during the immediate aftermath. Lastly, in the case that all normal lines of communication are down, including cellular data service, keeping a battery-powered radio in the building will allow you to at least receive updates from emergency response teams outside the disaster area.

The Federal Emergency Management Administration (FEMA) is an excellent resource to consult when building a business emergency response plan and crisis communications policy. For detailed guidelines on crisis communications, visit: http://www.ready.gov/business/implementation/crisis.

6.8.2 Tornadoes

Locate the best available source of accurate weather information, such as a local news station, the state police, or a state emergency agency. Weather band radios are useful in case other communications are interrupted. Locate suitable tornado shelters in the facility, preferably in a southwest corner below ground level in an area free of glass. If you lack such an area, instruct employees to avoid areas with exposed glass and to take shelter under heavy furniture or in reinforced areas in the center of the facility. Instruct employees to move to these designated areas by using prearranged signals or an announcement over the public address system.

If a tornado is imminent, begin emergency shutdown procedures, including gas mains and power supplies. Monitors should make sure that everyone has taken cover and that sufficient windows are open to lessen the dangers of explosion or implosion. Because tornadoes can create vacuum conditions, first-aid personnel should be alert for individuals suffering from loss of breath. If a facility suffers tornado damage, employees should be evacuated as soon as it is safe to leave shelter. Caution employees against touching metal objects or fallen wires. Do not allow employees to re-enter a damaged facility until it has been determined the structure is safe.

6.8.3 Winter Storms

Facilities located in areas subject to severe winter weather should prepare in advance for storm conditions that may hamper operations. You may want to define conditions under which employees should be dismissed early or advised to remain at home. In the latter case, establish a system for announcing facility closings.

Arrange for a skeleton crew in case severe blizzards cause a facility closing. This may require advance training of selected employees to carry out a number of different functions.

Arrange for snow removal and emergency auto service well before the onset of winter. It may also be advisable to stockpile snow fencing, salt, and sand early in the year. In remote locations with a history of sudden severe storms, consider providing sleeping quarters in case travel becomes inadvisable. Publish winter storm safety rules in newsletters or post this information on bulletin boards.

6.8.4 Floods

Vulnerability to flood conditions takes into account your area's flood history, the facility's elevation, and whether the facility is in a flood plain. You can contact the nearest office of the US Army Corps of Engineers (http://www.usace.army.mil/) for flood control assistance and information on potential flood conditions. Planning for floods should include these steps:

- Determine whether furnishings, equipment, materials, and supplies can be stored safely above ground level within the facility.
- Determine how to protect machinery that can't be moved.
- Select high-ground storage locations if needed.
- Find sources of packing materials in case you need to remove the contents of the facility.
- Assess the feasibility of finding an alternate operating facility.
- Plan how to secure the facility prior to the onset of flooding. People assigned to this task should make sure gas mains and power sources are shut down, and that doors and windows are locked and bolted.

6.8.5 Earthquakes

Managers of facilities in quake-prone areas must be certain to instruct employees to respond properly in the event of an earthquake. Instructions for employees should include the following:

- Stay inside and take cover under heavy furniture.
- Stay as close to the center of the facility as possible.
- Stay away from glass windows and doors.

- Avoid places where there is danger of falling debris. The greatest danger in earthquake conditions is from falling material.

When the tremor has ceased, employees should be taken to evacuation areas until damage to the facility structures is assessed. Establish a system to account for all employees and to provide assistance to the injured. If damage has occurred, assign personnel to check utilities and shut off power, gas, and electricity. If gas leakage is detected, open the windows and vacate the facility until utility representatives determine it is safe.

6.8.6 Hurricanes and Typhoons

Protecting employees and property from high winds and subsequent flooding is the primary concern with hurricanes. See the earlier sections on tornadoes and floods for planning information.

The National Hurricane Tracking Center in Coral Gables, Florida (http://www.nhc.noaa.gov/), provides the latest information and emergency procedures in the event of a hurricane. For up-to-date storm surge information, live updates about hurricanes and tropical storm events can be found on Twitter via @NHC_Surge.

6.8.7 Fires

Managers of facilities that are located in areas where there is a possibility of forest or grass fires should consider these planning factors:

- Keep the facility perimeter clear of brush, debris, and fallen trees.
- Establish fire trails around the facility to prevent any fires from spreading to company property.
- Plan for the removal of solvents and other flammables when forest fires are imminent.

6.9 STRIKES AND LABOR DISTURBANCES

Your first step in planning for strikes and labor disturbances is to monitor the expiration dates of labor contracts. Contact public relations when any strike deadline is near, and notify employee relations and human resources about any situations they may not be aware of.

The primary function of security during a strike is to protect people and the facility. Your plans should cover:

- Coordinating law enforcement assistance;
- Coordinating security efforts in emergencies;
- Controlling access;
- Protecting the premises;
- And establishing communications and other logistical considerations.

Of special concern, it's possible that strikes and labor disturbances can result in various problems for customers and products, such as business interruptions, shipment delays, tampered or damaged products, inventory buildup, and even customer harassment.

6.9.1 Law Enforcement

Prior to any strike deadline, you should meet with local police officials to determine under what circumstances they can help you maintain security. Remember, it is not their role to be strike breakers or contract negotiators. In some jurisdictions, police will act only to enforce a restraining order or injunction. Define the circumstances under which you will ask for police assistance, and the nonserious situations in which you will not. When you request police intervention for handling illegal activity during a strike, facility management should agree beforehand to cooperate with any prosecuting actions that may result. Your failure to cooperate in this would reflect poorly on the company and could result in poor police response when you need assistance in the future.

6.9.2 Coordinating Security in Emergencies

During a strike, be prepared for fire, medical, and other emergencies. Because of the chance of sabotage and the possibility that assistance may be hampered during a strike, it is a good practice to evaluate your facility's medical assistance and fire protection systems before a strike deadline.

- Inspect all fire extinguishers and auxiliary firefighting equipment.
- Arrange for extended fire patrols.
- Check the first aid station for adequate supplies.
- Arrange to have employees trained in first aid or paramedics available on short notice.
- Arrange for ambulance service before the strike, and notify local emergency medical facilities of the impending strike.

6.9.3 Controlling Facility Access

Find out how many non-striking employees or temporary workers will be entering the affected facility (this will depend on management decisions about carrying on with normal or limited operations). Then decide on the fewest access points needed for this level of operation. These access points should be in easily monitored areas and provide the quickest and most direct entrance into the facility. Keep picket line crossings to a minimum by encouraging employees to use car pools and discouraging them from leaving the facility for lunch. Consider eliminating unnecessary shipments and reducing others to a minimum.

During the strike, security personnel should cover the access points, and report and record any incidents or pickets that block movement or damage incoming or outgoing vehicles.

If possible, establish a buffer zone of 20 to 30 yards between the picket line and the facility perimeter. This increases the distance between pickets and possible targets of vandalism, creates an area where movement can be easily observed, and establishes a definite line of demarcation.

6.9.4 Site Protection

Site protection functions may need to be expanded during a strike. Determine whether you will need additional security personnel. If your facility uses contract security services, you should determine whether additional security officers will be available (note: some states prohibit agencies from providing guards solely for strike situations). Locations with proprietary security personnel, or with no security officers, may have to assign security duties to other employees.

Instruct people assigned to site protection functions that their duties are only to detect and report abnormal conditions. They should not attempt to enforce the law or confront violators. Caution them against discussing strike-related issues with the picketers.

The placement of protection personnel will vary with each location, but suggested areas include each entrance point and critical areas that are susceptible to vandalism.

6.9.5 Communications Center

Communications are your most important logistical consideration in planning for a strike. Establish a communications center with adequate radio communications between the various observation points. You will also need to establish communication channels with police, fire, medical, and other vital services.

The communications center should be the receiving point for all incoming information and the originating point for all instructions. Equip the center to monitor all radio traffic, and have one or more dedicated telephone lines. The center should

- Maintain a log of all abnormal incidents;
- Monitor radio traffic;
- Maintain contact with police and other vital services;
- Collect and process evidence obtained by observation teams; and
- Monitor local media broadcasts.

Individuals with decision-making authority should be in the center at all times to

- Coordinate security efforts;
- Establish observation teams;
- Arrange for the safe movement of people;
- Summon police and fire assistance; and
- Make required emergency decisions.

Observation teams equipped with cameras or other recording equipment can document activities that seem to go beyond normal picketing activity. These teams should

- Contact the communications center to talk to the decision-making authority to determine the legality of activities they are observing;
- Carefully record the date, time, and location of any incidents; and
- Objectively report what they observe.

It is important that you closely monitor security situations after a strike, and that you promptly report any incidents.

6.10 MEDICAL EMERGENCIES

Security personnel and managers with security responsibilities are often the first ones to be notified when a medical emergency occurs. Planning for medical emergencies will help you deal with both isolated incidents and those that result from a larger emergency.

You should establish extra safety procedures for those people who work alone or in isolated areas, or those with potentially serious medical conditions. You can arrange to have other staff check on these people periodically, and special alarms or call-in devices can also be given to these people so that help can be summoned quickly.

6.10.1 First Aid

You should have a sufficient number of people trained in basic first aid and cardiopulmonary resuscitation (CPR) so that you can quickly help accident victims and render aid in large-scale emergencies. You should arrange frequent refresher training for these people as part of their annual professional development program.

Establish a system for calling trained responders in an emergency. This can be a paging system, a public address system, or a telephone alert system.

Designate and equip a first aid area to handle common injuries. In case of an accident or disaster involving multiple injuries, you should also have a larger area where the injured can be taken until further help arrives.

6.10.2 Other Resources

Advance planning should include a survey of local medical facilities to determine which ones are best equipped to handle certain emergencies such as burns, cardiac care, and ingestion of toxic substances.

You should also determine the availability and quality of transportation services in your area, and identify alternate means of transportation if normal modes become unavailable.

CHAPTER 7

Special Security Situations

This chapter includes information about circumstances that pose special security considerations:

- The use of contractors, vendors, temporary employees, and interns
- On-site construction work
- Vacant company property awaiting sale, transfer, or development
- Special events

7.1 CONTRACTORS, VENDORS, TEMPORARY EMPLOYEES, AND INTERNS

There are many situations that involve the use of contractors, vendors, temporary employees (temps), and interns. These situations might require the presence of many such people on site, or perhaps only a few but over an extended period.

Most of these security considerations are not out of the ordinary, but the risks should be carefully assessed. In addition, the legal, human resources, safety, and security concerns should be reviewed with the appropriate company departments. These concerns and some special precautions are outlined below.

7.1.1 Access Controls

Access to your facility and company assets by contractors, vendors, temps, and interns should be strictly controlled.

- Parking areas should provide no greater vehicle access than that for employees unless required by contract.
- They should use only designated entrances to the facility, sign in and out at the designated area, and be able to provide proof of their company identity.
- They should wear the appropriate credentials while at the facility, and may need to be escorted to the work area in some cases.

7.1.2 PreEmployment Issues

The employment background of the contractor, vendor, temp, or intern you're hiring is an important consideration. For example, past or future work with a business competitor may need to be reviewed, and a background check may be necessary if their work involves very sensitive areas. Nondisclosure agreements should be signed if applicable.

7.1.3 Orientation and Awareness

You should always provide some kind of facility orientation so that the contractors, vendors, temps, and interns your company has hired have an understanding of the security protocols you've implemented. This orientation should include the following topics:

- When and where they work
- The areas they have access to, and those that are restricted
- What equipment or other property they may use or take off-site
- Parking restrictions and vehicle use
- Building access controls and procedures
- Information security procedures and controls, including what information or computers they have access to, and their responsibility to keep company information secure
- Confidentiality about company information, including the need to verify who they are talking to on phone calls
- The internal company contacts and resources they may need
- The emergency procedures they should follow
- How to report incidents and problems

Please note: It's possible that your company's legal counsel may also need to evaluate contracts when contractors' work requires access to proprietary information, equipment, procedures, or processes.

7.1.4 Service Vendors

Sometimes, service vendors may need to make a visit to your facility to perform work on-site. For example, equipment installation and maintenance, such as telecommunications systems, require service calls by technicians. In addition to the security considerations listed above, the following points are also recommended:

- Service vendors should inform the appropriate administrator of the results of the service call and whether another call is necessary.

- A company employee should monitor the service vendor when work is being performed after normal business hours.

7.2 CONSTRUCTION SECURITY

Construction sites require many security considerations for obvious reasons. You should appoint a company representative to be the point of contact for a construction contractor. This representative should brief the contractor on all applicable security and safety procedures, access controls, company policies, and other requirements or restrictions.

7.2.1 Security Measures

The following security measures are frequently used in construction situations:

- Additional access control systems and procedures
- Padlock system and key disbursement controls
- Area warning and restriction signs
- Additional security officer staffing
- Contractor sign-in procedures
- Inspection notification and procedures
- Tool control procedures

7.2.2 Facility Access

Construction workers should have a designated fenced parking area where they are required to park their vehicles during the workday. Vehicles in violation of this regulation may be towed at the owner's risk and expense. All pedestrian access to the work site should be controlled, and it is recommended that construction workers have passes. Additionally, signs should clearly mark all restricted areas and entry points.

No management or hourly construction workers should be allowed to enter existing facility buildings except as agreed upon by your company's facility security coordinator and construction management. Areas that may be specifically restricted include

- Bathrooms;
- Lunchrooms;
- Warehousing areas; and

- Any area of the facility where the worker does not need to be for an authorized, work-related reason and where a visit has not been specifically authorized by the facility security coordinator.

All vehicles entering or leaving the construction site are subject to inspection. You may want to issue authorization decals to construction vehicles. Private vehicles should not be allowed to enter the work site except to make a specific pickup or delivery, and prior authorization for such entry should be required.

7.2.3 Shared Access

In some cases access to the site and buildings may have to be shared by employees and construction workers. In these cases, you may have to increase security efforts or adopt temporary measures to control access.

7.2.4 Construction Site Layout

The layout of the construction site may pose specific security concerns for controlling access to and movement within the site perimeter and building perimeter. The following areas may require special attention before construction begins:

- Temporary fencing and gates
- Parking areas for construction workers, trailers, and construction vehicles. These areas should be separate and should not provide greater vehicle access than that for employees unless required by contract.
- The construction office
- The security officer post

7.2.5 Construction Preparation

Once construction begins, there are several other security concerns to consider:

- Excavation notification and utility locations
- Construction photographs (in regard to information security and camera policy)
- Receiving construction materials
- Mail distribution

7.2.6 Removing Materials from the Work Site

Although your company is not responsible for theft or damage to construction equipment, materials, or tools, you might want to require a construction manager or supervisor to complete and sign a material pass whenever items that are clearly valuable are being removed from the work site. Construction vehicles carrying miscellaneous items do not usually need a material pass, but you might require one for items such as the following:

- Gang boxes
- Trailers
- Equipment larger than what is normally carried by hand
- Obviously new lumber
- Handheld power tools

7.2.7 Company Policies

Inform construction workers that they are responsible for knowing and complying with company policies prohibiting the possession or use of the following in construction offices or on the work site:

- Alcohol
- Beverages containing alcohol
- Containers normally used to hold alcohol
- Illegal drugs, including look-alikes
- Drug paraphernalia
- Cameras
- Weapons, firearms, and ammunition

7.2.8 Drug and Alcohol Notice

Post the following notice about your company's drug and alcohol policy on bulletin boards in construction offices and entrances to company property:

> **NOTICE**
>
> The possession or use of alcohol or beverages containing alcohol, containers that are normally used to hold alcohol, illegal drugs including look-alikes, drug paraphernalia, ammunition, firearms, or weapons are not allowed on these premises. All vehicles and personnel entering this facility/work site are subject to inspection.

The use of alcohol or other drugs can seriously impair a worker's ability to perform his or her job satisfactorily, and can frequently jeopardize the safety of employees and other people. Construction workers and their personal effects, including personal vehicles parked on company property, are subject to inspection for the items listed above. See Chapter 5, Inspection Guidelines, for more information.

7.3 VACANT PROPERTIES

Special security considerations apply to vacant company real estate properties awaiting sale, transfer, or development. The guidelines in this section will help you review and maintain overall security to protect against liabilities resulting from injuries to guests, trespassers, and animals.

7.3.1 Systems Maintenance

All security systems should be routinely inspected and all security procedures should be reviewed regularly. Validation testing should be completed and documented. The failure to identify and remove a risk can become a liability, and the failure to maintain or properly operate an existing system can create legal liabilities. Problems with lighting, locks, fencing, missing signs, and missing tank covers, for example, must be quickly identified and corrected.

7.3.2 Removing Hazards

Most common security risks can be minimized by using access controls, adequate warning signs, barriers and locking devices, and maintaining exterior lighting standards. Remove, secure, or inactivate operational and environmental hazards whenever possible, such as conveyors, presses, extruders, mixers, electrical junction boxes, transformers, storage tanks, chemicals, flammables, and standing water.

7.3.3 Facility Survey

After vacating or before assuming control of a facility, have the facility surveyed and photographed. Security concerns should be corrected and documented, and the survey and photographs should be added to the facility file.

7.3.4 Notification

Notify the appropriate company departments when you are managing a vacated facility. These departments may include: corporate security, corporate safety, fire protection, insurance, tax, facility engineering, mechanical, and telecommunications. Also obtain copies of all the service contracts pertaining to the site.

Notify and provide a contact person for the appropriate law enforcement agencies, local fire departments, public and private utilities, building inspectors, and any contracted alarm and guard companies.

7.3.5 Intrusion Detection

Maintain and monitor intrusion detection systems. Quickly detecting and responding to intruders goes a long way toward protecting the property. Response, notification, and follow-up procedures can be combined with the fire protection program.

7.3.6 Required Access

Give consideration to police officers, firefighters, and public utility personnel who may have a legitimate need and legal authority to enter the facility. Special locking devices or local property management personnel can usually serve this need. At a minimum, provide them with the name and telephone number of the company person they should contact.

7.3.7 Trespassing

Discourage trespassers and remove attractive nuisances (childhood curiosity about an unoccupied facility requires special attention). The company must also protect against criminal activity at the vacant property. Signs of drinking, drug use, vandalism, homeless people sleeping at the site, or regular access should be acted on immediately.

All facilities should also have the appropriate "No Trespassing" signs posted.

7.3.8 Local Ordinances

The considerations listed above should be reviewed and implemented in accordance with state statutes and local ordinances for vacant buildings. Because of differences between jurisdictions, each property should be reviewed individually.

7.3.9 Property Disposition

Notify contractors, local agencies, and the appropriate company departments of the final disposition of the property. Any proprietary security equipment, locks, and signs used during the transition should be recovered.

7.3.10 Inspection Schedules

Schedule periodic inspections to monitor the property. These inspections should include the maintenance of alarm and other systems. If contractual security guard services are used, the contract, specified duties, and procedural instructions should be reviewed.

The checklist below can help you when you conduct periodic security or engineering inspections of vacant facilities.

- Inspect perimeter fencing. Fence fabric and wire strands should be in good condition, tightly stretched, with no gaps or cuts.
- Inspect the property and verify that it is signed in accordance with local ordinances. The signs should be in good condition, clearly stating "Company Property — No Trespassing."
- Inspect the perimeter lighting. Every fixture should be in working condition and the illumination level measured to make sure exterior lighting standards are being met.
- If the property is fenced, the gates should be in good working order with adequate locking devices. "No Trespassing" signs and emergency notification phone numbers should be posted.
- Any buildings on the property and all access points, including doors and windows, should be secured.
- Change all perimeter locks to ensure that only authorized individuals can access the facility. You may want to purchase a number of cylinders and key sets keyed the same for all managed properties, so that only one key is needed to conduct all inspections.
- Inspect the fire alarm/suppression and intrusion alarm systems. Make sure they comply with the recommended inspection and testing schedule. Be certain that the phone lines used to transmit fire and intrusion alarm signals are retained and in good operating condition. Get in touch with the central monitoring station personnel and give them a new list of the appropriate people to contact.
- Inspect and secure all electrical rooms and substations.
- Inspect all tank and underground storage facilities, and secure access points if necessary.
- Inspect the facility for hazardous materials or substances. Remove or secure any such materials, taking proper safety precautions.

- Notify local agencies of building status and emergency notification procedures.

7.4 SPECIAL EVENTS

Special events on company property can present unique security considerations. As soon as you learn about an upcoming special event, you'll need to assess the risks and identify the role of security, both your team and corporate (if applicable), as well as the public and private security resources available.

Special events can include:

- Open houses
- Group tours
- Special on-site meetings or executive visits
- Visits by groups representing particular issues or concerns (for example, environmental issues or animal rights)
- Use of company facilities by nonemployees, such as community groups
- Visits by high-ranking government officials, such as the president, vice president, and cabinet level officers, or by controversial people
- Sponsorships of major events, such as for the Olympics, tournaments, other sporting events, and charities
- Telethons and visits by celebrities

CHAPTER 8

Security Management Resources

The security industry continues to change and develop. To stay current with these changes, use the many resources available to you:

- Training programs
- Professional associations
- Trade publications

This chapter recommends some resources in each of these areas. ASIS International (www.asisonline.org) is another excellent resource for those in the security management field. On the ASIS website, you can find helpful information about educational programs and workshops, conferences, and the Certified Protection Professional program.

If your company has a corporate security department, it too should serve as an excellent resource of literature and advice for your security management needs.

8.1 TRAINING AND EDUCATION

There are two types of training that can help you build a successful security program: awareness education for employees and direct training for the individuals responsible for security.

Employees must be aware of the security measures at the workplace and what their security responsibilities are.

Direct training ensures that individuals responsible for security will stay on top of the many changes in the industry, and that they will be better able to participate in the security program.

In addition to formal training programs, there are many conferences and seminars throughout the country on security issues. These provide good opportunities for you to meet with other business-people concerned with security issues.

Here's a sample of specific conference topics:

- Designing Facility Security Systems
- National Security and Risk Management Conference
- Physical Security Workshop
- Crime Prevention Theory, Practice, and Management
- Disaster Recovery and Contingency Planning
- Loss Prevention

8.2 PROFESSIONAL ASSOCIATIONS

Professional associations are excellent sources of security information, and some of them have local chapters that you can join. Below is a partial list of associations, but contact your local law enforcement liaison for information about local chapters.

- Computer Security Institute: Provides information on security data, publishes the *Computer Security Journal,* and offers seminars and training.
  http://www.gocsi.com/
- Door and Hardware Institute: Distributors of doors and builders' hardware; architectural consultants.
  http://www.dhi.org/
- Institute of Internal Auditors
  http://www.theiia.org/
- International Association of Chiefs of Police
  http://www.theiacp.org/
- International Society of Crime Prevention Practitioners
  http://www.iscpp.org/
- ASIS International
  http://www.asisonline.org/

8.3 TRADE PUBLICATIONS

There are many types of security trade publications that will keep you up to date, including newsletters, journals, books, and magazines. Some of the more popular are listed below.

8.3.1 Newsletters

- For a comprehensive list of security newsletters, visit:
  http://www.mainesecurity.com/Security_Newsletters.htm

8.3.2 Journals

- The *Security Journal*
  http://www.palgrave-journals.com/sj/index.html

8.3.3 Magazines

- *Security Management*
  http://www.securitymanagement.com/
- *Security*
  http://www.securitymagazine.com/

8.4 BOOKS

Butterworth-Heinemann, an imprint of Elsevier, has an extensive collection of security, risk management, and emergency management books. For a full list of available titles from this and other Elsevier security imprints, visit: http://store.elsevier.com/Security/C_2/.

For other titles in Elsevier's Security Executive Council Risk Management Portfolio, including other playbooks, visit: http://store.elsevier.com/SecurityExecutiveCouncil.

CHAPTER 9

Implementing Your Security Program

This chapter summarizes the key takeaways from the previous eight chapters, and can serve as a model for building your own reference source of information about your specific contacts and local procedures. You can use this chapter as a template for your security program, or you can use these ideas as a starting point for putting together a companion guideline to your security program

If you have already put together your security program, use this chapter as a checklist to see that you have covered all the major points.

Update your documentation on a regular basis. It is an important part of your security program, and an important tool for assessing your security needs.

9.1 COORDINATORS

The most important part of organizing a security program is getting everyone involved. Some people must take special responsibility for filling critical roles. Keep on file the names of and specific responsibilities for the following individuals:

- Facility security coordinator
- Information security coordinator
- Liaison person with law enforcement

9.2 SECURITY COMMITTEE

Record the names and responsibilities of all members of your security committee. You should also keep a record of meeting agendas and minutes and the outcome of any specific issues that have been discussed.

9.3 EMPLOYEE AWARENESS

Maintain a record of all employee awareness efforts. These can include

- Presentations;
- Articles published in employee newsletters;
- Bulletin board announcements; and
- Training programs.

You should also maintain a copy of all your company policies and guidelines on employee issues and concerns (and those of your corporate security department, if applicable), as well as any specific rules or procedures you have established in these areas. These areas may included the following:

- Cameras
- Firearms
- Alcohol and drugs
- Gambling and lotteries
- Parking and auto registration
- Securing valuables

9.4 INFORMATION SECURITY

Put together a guideline that contains all relevant information about your information security program, its coordinator, your document classification scheme, your data processing safeguards and passwords, and the other controls you've established. Update the guideline regularly to ensure that it stays current.

9.5 ACCESS CONTROLS

It is extremely important that you control all entrances to and exits from your facility. In order to do this, you will need to consider the items in this section.

9.5.1 Identification

Determine the types of badges or other identification required for

- Employees;
- Visiting company employees;
- Temporary employees;

- Customers;
- Visitors (including tours, families, and applicants);
- Contractors; and
- Trade representatives.

9.5.2 Procedures

Set up procedures for:

- General sign-in/sign-out practices;
- After-hours access;
- Notification and authorization for any special access request;
- Escort policy;
- Shipping and receiving (dock and truck controls);
- And property passes.

Be certain that you identify the special passes and procedures required for removal of company property from the premises, including

- Material passes;
- Surplus sales receipts; and
- Contractor equipment passes.

9.5.3 Key Control

Designate a key custodian who is responsible for issuing and tracking all building keys. Control the levels of master keys and the number of duplicates. Establish strict procedures for key control. You should also keep copies of any requests for keys and copies of key receipts. Identify any special procedures for issuing and tracking keys for office desks, file cabinets, supply cabinets, etc., and establish procedures for reporting, turning in, and claiming lost keys (both company and personal keys).

9.5.4 Automated Access Control

An automated access control system begins with its design and installation. After that, sound management of the system determines its success. The list below can help you manage your system.

- The goal of the system is “no alarms” (forced doors, doors held open, etc.), which means that security procedures are being followed.

- The system requires continuous monitoring; therefore, terminals and printers must be located in an area where they are used regularly.
- The entry of new cards and deletion of old ones must be kept current. It is important that lost, stolen, or nonfunctioning cards are removed quickly from the system.
- Exception reports must be run on a regular basis.
- The system must be routinely maintained by the appropriate service vendor.
- It is the responsibility of those monitoring the system to observe all actions and to report all exceptions to the appropriate management person.

9.5.5 Alarm Systems

Alarm systems must also be monitored closely to be successful. There is no such thing as a false alarm—there are only unexplained alarms. The following list provides a few key points for managing an alarm system.

- All employees working after normal business hours should have unique entrance codes. You need to be able to tell which individual used the facility after hours, on a case-by-case basis.
- Keep the list of people authorized for after-hours access as short as possible.
- At least once every month, get a report of the exceptions to normal openings and closings.
- Get an alarm history report once each month.
- Investigate all alarms to ensure that you know the reason each alarm was given.

9.6 SECURITY OFFICERS

Security officers have duties and responsibilities that vary widely from company to company, depending on whether they are trained security professionals, contractors, or regular company employees that share security responsibilities. If you do hire contract security services, be aware that their contractual arrangement and legal jurisdiction may limit the responsibilities of their officers.

If you do not already have a handbook for your security officers—whoever they may be—use this section to define their roles and responsibilities.

The primary function of your security officers is to observe and report. Response is not their main function. In addition, it is important to remember that security officers are often a person's first contact with your company. Because of this, you should carefully define any expectations you have about general hospitality and public relations.

9.6.1 General Responsibilities

Security officers must be generally aware of company policies, facility procedures, and the layout of the physical site. Their knowledge should include

- Company policies and safety rules;
- Your site and building layout, including the location of all exits, alarm systems, hazards, and special equipment;
- Safety and fire regulations; and
- Emergency contacts and procedures.

9.6.2 Specific Duties

At a minimum, security officers should be responsible for the following:

- Controlling employee and visitor access
- Contacting emergency help
- Patrolling and inspecting the site at designated intervals
- Verifying that all security equipment is accounted for and working properly
- Reporting all security-related incidents and concerns

9.6.3 Procedures for Security Rounds

Establish regular security rounds and reporting procedures. Be certain to include

- Site and building perimeter rounds;
- Building interior rounds; and
- Other checks and inspections that are appropriate to your site.

9.7 SECURITY SURVEYS

9.7.1 Surveys and Validations

You should have copies of the following surveys and validations:

- Facility profile (most recent)

- Validation tests and results
- The most recent corporate security survey (if applicable) and your response to its assessment and recommendations

9.8 REPORTING

9.8.1 Forms

Compile all forms and procedures that you use for reporting security incidents.

9.8.2 Contacts

List all the people who should be notified about security incidents. See the section on "Emergency Plans and Contacts" later in this chapter for specific examples of internal and external contacts.

9.8.3 Reports

Have copies of all security incident reports for the last 12-month period, including

- Local incident reports;
- Results of any follow-up to these reports;
- Distribution list and response requirements for all security reports and memos; and
- Any incident reports submitted to corporate security (if applicable).

9.9 INVESTIGATIONS AND INSPECTIONS

9.9.1 Investigations

Maintain current contact information on

- Local investigative assistance;
- Local management notification requirements;
- Divisional notification requirements; and
- Corporate notification requirements.

9.9.2 Inspections

Maintain complete documentation on any inspections that have been conducted.

9.10 EMERGENCY PLANS AND CONTACTS

9.10.1 Plans

You should have plans in place for all probable emergencies and natural disasters. Review and update the following plans and documents:

- Site plan
- Evacuation route and procedures
- Demonstration plan
- Bomb threat plans and procedures
- Natural disasters: safety, security, and communications plans

9.10.2 Organization

You should have a team in place that is ready to provide support for the public safety agencies that respond to emergencies on company property. The team should include individuals with your company who represent the following functions:

- Administration
- Security
- Safety
- Fire systems maintenance
- Medical services/industrial hygiene/health physics
- Environmental engineering and pollution control
- Public relations/human resources

9.10.3 Company Emergency Contacts

List the names and phone numbers of the company employees who you must contact first in the event of an emergency. You should have also their home phone numbers and a list of alternates to guarantee a 24-hour response. These employees include:

- Facility security coordinator
- Emergency squad members
- Members of the company's safety department
- Management
- Corporate security (if applicable)

Additionally, you should list the department names and phone numbers of all external agencies you might need to contact during an emergency, including

- The local police department;
- Sheriff's office;
- Fire department;
- Highway patrol; and
- Alarm response (central station).

List all medical help that you can contact in the event of sudden injury or illness. In addition to community hospitals and clinics, it is important that you keep an updated list of company employees who can provide medical assistance, such as people who are trained in first aid and CPR. This list will include

- On-site medical services department;
- On-site first aid trainees;
- Hospitals (note areas of specialized emergency assistance);
- Clinics;
- Doctors (clinic and residence); and
- Ambulance service.

Keep an updated list of vendors who service the following security systems and who can provide information and assistance in the event of an emergency:

- Lock and key systems
- Alarm systems
- Automated control systems

Also keep an updated list of people who service the following building systems when they malfunction or breakdown, and who can also provide information and assistance in the event of an emergency.

- General equipment and building problems
- Electrical equipment
- Plumbing
- HVAC
- Mechanical

Lastly, also maintain a list of the company managers and engineers (listed below) who can provide information about the physical plant in case of emergencies.

- Facility manager
- Facility engineer
- Office manager

9.11 SPECIAL SECURITY REQUIREMENTS

Keep a record of all circumstances that may require special security considerations. These are outlined in Chapter 7, and may include

- Visits from contractors, vendors, temporary employees, and interns;
- New construction;
- Vacant properties;
- Special events; and
- Acquisitions and divestitures.

9.11.1 Local Resources

Keep a record of all your internal and external resources. These may include

- Professional associations you belong to;
- Training and education sources you rely on, such as videos, films, and training programs; and
- Any publications that you use.

About Contributing Editor

Bob Hayes has more than 25 years of experience developing security programs and providing security services for corporations, including eight years as the CSO at Georgia Pacific and nine years as security operations manager at 3 M. His security experience spans the manufacturing, distribution, research and development, and consumer products industries as well as national critical infrastructure organizations. Additionally, he has more than 10 years of successful law enforcement and training experience in Florida and Michigan. Bob is a recognized innovator in the security field and was named as one of the "25 Most Influential People in the Security Industry" in 2007 by *Security* magazine. He is a frequent speaker at key industry events. He is a leading expert on security issues and has been quoted by such major media outlets as the *Wall Street Journal* and *Forbes*. Bob is currently the managing director of the Security Executive Council.

About Elsevier's Security Executive Council Risk Management Portfolio

Elsevier's Security Executive Council Risk Management Portfolio is the voice of the security leader. It equips executives, practitioners, and educators with research-based, proven information and practical solutions for successful security and risk management programs. This portfolio covers topics in the areas of risk mitigation and assessment, ideation and implementation, and professional development. It brings trusted operational research, risk management advice, tactics, and tools to business professionals. Previously available only to the Security Executive Council community, this content—covering corporate security, enterprise crisis management, global IT security, and more—provides real-world solutions and "how-to" applications. This portfolio enables business and security executives, security practitioners, and educators to implement new physical and digital risk management strategies and build successful security and risk management programs.

Elsevier's Security Executive Council Risk Management Portfolio is a key part of the **Elsevier Risk Management & Security Collection**. The collection provides a complete portfolio of titles for the business executive, practitioner, and educator by bringing together the best imprints in risk management, security leadership, digital forensics, IT security, physical security, homeland security, and emergency management: Syngress, which provides cutting-edge computer and information security material; Butterworth-Heinemann, the premier security, risk management, homeland security, and disaster-preparedness publisher; and Anderson Publishing, a leader in criminal justice publishing for more than 40 years. These imprints, along with the addition of Security Executive Council content, bring the work of highly regarded authors into one prestigious, complete collection.

The Security Executive Council (www.securityexecutivecouncil.com) is a leading problem-solving research and services organization focused on helping businesses build value while improving their ability to effectively manage and mitigate risk. Drawing on the collective knowledge of a large community of successful security practitioners, experts, and strategic alliance partners, the Council develops strategy and insight and identifies proven practices that cannot be found anywhere else. Their research, services, and tools are focused on protecting people, brand, information, physical assets, and the bottom line.

Elsevier (www.elsevier.com) is an international multimedia publishing company that provides world-class information and innovative solutions tools. It is part of Reed Elsevier, a world-leading provider of professional information solutions in the science, medical, risk, legal, and business sectors.